A PROJECT MANAGER'S BOOK OF TEMPLATES

A PROJECT MANAGER'S BOOK OF TEMPLATES

Cynthia Snyder Dionisio

WILEY

Published by John Wiley & Sons, Inc., Hoboken, New Jersey.
Published simultaneously in Canada.

For general information on our other products and services or for technical support, please contact our Customer Care Department within the United States at (800) 762-2974, outside the United States at (317) 572-3993 or fax (317) 572-4002.

Wiley also publishes its books in a variety of electronic formats. Some content that appears in print may not be available in electronic formats. For more information about Wiley products, visit our web site at www.wiley.com.

Library of Congress Cataloging-in-Publication Data is Applied for:
Paperback ISBN: 9781119864509

Cover image: © Sigit Mulyo Utomo/Getty Images
Cover design: Wiley
SKY10046345_042123

Contents

Acknowledgments

My gratitude to all the professors and teachers who use this book as a resource. I am so glad you find this a useful reference for your students. For all the students who use this book, I hope it makes learning the art, science, and magic of managing projects a bit easier. For professional project managers, I hope you find some value in the hybrid approach that is incorporated in this book. For the accidental or occasional project manager, I am so glad you have found this book. I hope it will reduce the time you spend figuring out how to organize, lead, and document your project, so you can focus on getting the work done!

Thank you to Kalli Schultea and Amy Odum for shepherding this book through the publication process. I feel so fortunate to work with you and all the fabulous professionals at Wiley.

About the Companion Website

The companion website can be found at

www.wiley/com/go/dionisio/bookoftemplates

Introduction

The project management profession is evolving rapidly. We have moved from methods that were purely predictive (aka waterfall development) or adaptive (Agile), to a more blended or hybrid approach. In fact, today more than 50% of projects are practicing formal or informal hybrid project management.

A Project Manager's Book of Templates provides templates that address project artifacts for predictive management, such as a schedule management plan and a lessons learned register; as well as templates for adaptive and Agile project management, such as a requirements backlog and release plan.

Another shift in the project management profession is that more and more often we are accountable for project and business results. To help you with that, you will find business templates for a business case, startup canvas, project proposal, and other templates to help you present business and strategy level information to senior level stakeholders.

AUDIENCE

This book is written for project managers in traditional fields, such as construction and system implementation, as well as project managers in the digital domain, such as software development, digital product management, and high-tech. Because the book is tactical rather than theoretical, it can be used by novice and advanced practitioners, academia, test prep, and as an on-demand reference.

Those new to project management can use the templates as a guide in collecting and organizing project information. Experienced project managers can use the forms as a template so that they collect a set of consistent data on all projects. In essence, these templates save reinventing the wheel for each project.

A secondary audience is the manager of project managers, a project management office, and program managers. Using the information in this book ensures a consistent approach to project documentation. Adopting these forms on an organizational level will enable a repeatable approach to project management.

This book does not teach project management concepts or describe how to apply project management techniques. Textbooks and classes can fulfill those needs. This book provides an easy way to apply good and consistent practices to projects.

ORGANIZATION

Content in this book is structured to align with the type of work we do on projects. For example, we do some planning, we track information with registers, and we have lots of project documents to help us stay organized. These are the sections in this book:

Starting the Project. Templates in this section are used to address business needs, provide high-level information, and authorize the project. Examples include the project charter and a vision statement.
Project Plans. Templates in this section support developing plans that will be used to guide project delivery. You will find templates for a risk management plan, release plan, change management plan, and many others.

Project Documents. Project documents help us develop and organize information we need for the project. There are many document templates, including estimating worksheets, requirements traceability matrix, and user stories.

Logs and Registers. Logs and registers are updated throughout the project. They help us keep track of dynamic aspects of our projects. You will find templates such as a backlog, stakeholder register, and assumption log.

Reports and Audits. To help track and report on the project we can use various project reports and conduct audits on specific aspects of the project. In this section, you will find a procurement audit, quality report, and project closeout report, among others.

Each template includes a description of the elements included in the template and a sample form that shows how those elements may be arranged. Every template should be tailored and modified to meet your needs. The description and sample templates are here to provide you with ideas to help you collect and manage the information needed to make your project a success.

Most template descriptions follow this format:

- A description of the template is presented along with a list of contents. For the planning forms, there is a description of where the information comes from and where it goes to.
- A section that presents information you can consider for tailoring the templates to fit your needs.
- An alignment section that presents related templates that you will want to make sure are aligned.
- A description table that identifies each of the fields in the template along with a brief explanation.
- A blank copy of the template.

I have also included an appendix that has some samples of combined templates. In the tailoring section, I have made some suggestions on how you can tailor a template by combining it with another. The appendix does not talk you through the details for the templates because the information is available where the templates are first discussed.

As I am sure you know, not all templates will be needed on all projects. Use the templates you need, to the degree that you need them. I hope you find value in the templates I have included in this book.

Starting the Project

There is no doubt that starting a project off right is the first step in delivering a successful project. Because projects vary greatly in size, methodology, criticality, and stakeholders, there are several ways you can compile and document the initial project information.

The templates in this section document high-level information that is later elaborated in project plans and project documents. Templates that are commonly used to document the initial project information include

- Project proposal
- Business case
- Project startup canvas
- Vision statement
- Project charter
- Project brief
- Project roadmap

The project proposal, business case, and project startup canvas are usually developed prior to a project being formally approved. They have information that helps relevant stakeholders determine if the need for and benefits of the project justify the investment of time, budget, and resources. These templates may be developed by a project sponsor because the project has not yet been approved and a project manager has not been identified.

The vision statement, project charter, and project brief templates are typically completed once a project has been approved. They provide a high-level view of the project. They may be developed by the project sponsor, the project manager, or by both of them working together.

A project roadmap takes information from the previous documents and creates a summary level graphic display of information. It is developed by the project manager.

Most projects are good with using two or three of these templates to get the project started. Much of the information in these templates is found in multiple templates. Therefore, you should determine the best template for your project and then tailor it to meet the needs of your project by editing, combining, or revising the template.

Project startup templates are usually developed once, before, or shortly after the project is authorized. They provide information on the business environment, justification for the project, financial expectations, and high-level information about the project. If there is a significant change in the environment or the project, the need for the project may be revisited and these documents may be updated.

1.1 PROJECT PROPOSAL

The project proposal is a proposition that describes an opportunity, a solution to a problem, or an approach for undertaking a mandatory project. Ideally, it is no more than one or two pages. A project proposal provides information about the environment, why a project is needed, and presents the proposed response and approach for the project. It is used to provide high-level information so decision makers can determine if the project should be undertaken.

Typical information includes

- Executive summary
- Project background
- Solutions and approach
- Financial information
- Resource requirements
- Conclusion

It may provide information to

- Business plan
- Project startup canvas
- Project charter
- Project brief
- Project roadmap

It is developed once, and then only changed if there are significant changes in the market, the environment, or the need.

Tailoring Tips

Consider the following tips to help tailor the project brief to meet your needs:

- For new product development projects, you can combine the vision statement with the project proposal.
- For smaller projects, the project proposal and business case may be combined.
- If there is relevant research or studies, this information can be included in an appendix.

For hybrid projects, you may include information on methodologies that will be used to deliver effectively.

Alignment

The project brief should be aligned and consistent with the following documents:

- Business case
- Vision statement
- Project startup canvas
- Project charter
- Project brief
- Project roadmap

Description

You can use the element descriptions in Table 1.1 to assist you in developing a project proposal.

TABLE 1.1   Elements of a Project Proposal

Document Element	Description
Executive summary	A succinct overview of the problem or opportunity the proposed project will address along with the ways it will address it. Includes a synopsis of the background, project objectives, and deliverables.
Project background	Information that provides context for the project. May include history, environmental considerations, market conditions, significant events, or other information that shows a compelling need for the project.
Solutions and approach	A summary of the goals and scope of the project, the expected timeline for delivery and a brief description of the methodology that will be used to deliver the project.
Financial information	High-level project funding requirements. May include financial metrics.
Resource requirements	Brief description of the physical resources required, including material, equipment, and sites. A summary of the skill sets and number of team members required.
Conclusion	A summary of the key points.

PROJECT PROPOSAL

Proposed Project Title: _____ Date: _____

Executive Summary:

Project Background:

Solution and Approach:

Goals	Scope

Financial Information

Resource Requirements

Physical Resources	Team Resources

Conclusion

1.2 BUSINESS CASE

The business case describes the business rationale for undertaking a project. It describes the current situation, future vision, threats, opportunities, costs, and benefits. A business case contains market information, financial metrics, and alternatives to consider.

Typical information includes

- Executive summary
- Background information
- Project objectives
- Project benefits
- Project definition
- Market assessment
- Alternatives analysis
- Financial analysis
- Risk overview
- Appendices

The project business case can receive information from

- Project proposal
- Vision statement

It may provide information to

- Project charter
- Project startup canvas
- Project brief
- Project management plan

It is developed once, and then only changed if there are significant changes to the market, financial analysis, or project definition.

Tailoring Tips

Consider the following tips to help tailor the business case to meet your needs:

- For large projects, the alternatives analysis may be a separate document.
- You can include information on project governance for large projects.
- Projects that will use a hybrid approach may want to include a section on project approach to define which aspects of the project will use a predictive approach, which will use an adaptive approach, and how they will integrate.

Alignment

The business case should be aligned and consistent with the following documents:

- Project proposal
- Project charter
- Project management plan

Description

You can use the element descriptions in Table 1.2 to assist you in developing a project business case.

TABLE 1.2   Elements of a Project Business Case

Document Element	Description
Executive summary	Provide a summary description of the business case. Give stakeholders a brief overview of the project.
Background information	Describe the environment and business context for the project. Identify the problem or opportunity. Document how the project aligns with the organization's strategic plan.
Project objectives	The measurable objectives that project intends to achieve.
Project benefits	Describe the intended benefits, such as gaining efficiencies, improving quality, increasing revenue, etc.
Project definition	Describe the key deliverables and the project boundaries. As appropriate, describe the approach to achieve the deliverables.
Market assessment	Provide an overview of the marketplace, including technology availability and legal, environmental, and competitor information.
Alternatives analysis	Describe the alternatives that have been considered and your recommended alternative. For each alternative, provide benefits, costs, and risks. Document how each alternative meets the need or solves the problem. If appropriate, include a feasibility analysis for each alternative.
Financial analysis	Calculate key financial indicators, such as net present value, return on Investment, cash flow, and life cycle cost.
Risk overview	Describe high-level project threats and opportunities along with the potential impacts.
Appendices	Attach supporting information such as spreadsheets, research, and references.

PROJECT BUSINESS CASE

Project Title: _____ Date: _____

Executive Summary

Background Information

Objectives	Success Criteria

Benefits

Project Definition

PROJECT BUSINESS CASE

Market Assessment

Technology availability	
Legal and regulatory	
Environmental	
Competitors	

Alternatives Analysis

Alternative	Benefits	Costs	Risks
1.			
2.			
3.			

Financial Analysis

Alternative	NPV	ROI	Cash Flow	Life Cycle Cost
1.				
2.				
3.				

Risk Overview

Appendices

1.3 PROJECT STARTUP CANVAS

The project startup canvas is a high-level visual summary of a project. It is a framework that allows you to capture new project information quickly and succinctly. It is modeled after the business startup canvas and the lean startup canvas. Like those models, the project startup canvas is one page. However, the project startup canvas focuses on key project information rather than market, competitor, and distribution information. The project startup canvas can include information on

- Problem or opportunity
- Solution or scope
- Key deliverables
- Value proposition
- Stakeholders
- Resources
- Costs
- Milestones
- Threats and constraints

The project startup canvas can receive information from

- Project proposal
- Vision statement

It may provide information to

- Project charter
- Project brief
- Work breakdown structure
- Backlog
- Resource requirements
- Cost estimates
- Schedule
- Risk register

The project startup canvas is developed once and is not usually changed unless there is a significant change in the environment, scope, schedule, budget, or resources.

Tailoring Tips

Consider the following tip to help you tailor the project startup canvas to meet your needs:

- If your project is for new product development, consider focusing on distribution channels, customer segments, and revenue streams rather than deliverables, resources, and milestones. While these elements are more business oriented, documenting that information at the start helps keep the project focused on the end users and the market.

Alignment

The project startup canvas should be aligned and consistent with the following documents:

- Project vision statement
- Project scope statement
- Work breakdown structure
- Stakeholder register
- Milestone schedule
- Cost estimates
- Risk register

Description

You can use the element descriptions in Table 1.3 to assist you in developing a project startup canvas.

TABLE 1.3   Elements of a Project Startup Canvas

Document Element	Description
Problem/Opportunity	Identify the problems the project will solve or the opportunities it will meet. Include contextual information if it provides insightful value.
Solution/Scope	Describe the suggested solution to the problems. Note the scope boundaries (what is in scope and what is out of scope).
Deliverables	List key project and product deliverables.
Value proposition	Describe why the project is needed and the value it will provide.
Stakeholders	List the key stakeholders including the customer and end user.
Resources	Identify the key physical resources and important skill sets needed.
Costs	Provide an initial cost estimate. Include fixed and variable costs, and project and maintenance costs.
Milestones	List the key milestones.
Threats/Constraints	Identify significant threats and constraints to the project.

PROJECT STARTUP CANVAS

Problem/Opportunity	Solution/Scope	Value Proposition	Customers	Costs
	Deliverables		Resources	
Milestones		Threats/Constraints		

1.4 PROJECT VISION STATEMENT

The project vision statement provides the future view of a product or service being developed. The project vision statement should be aspirational, yet achievable and realistic. It is developed at the very beginning of a project and is often an input to the business case.

The project vision includes at least

- Target customers
- Needs addressed
- Product or service attributes
- Key benefits

The project vision statement can receive information from

- Project proposal
- Project startup canvas

It may provide information to

- Project charter
- Project brief

The project vision statement is often used for projects that use an Agile methodology to develop digital products. It is developed once, at the beginning of the project.

Tailoring Tips

The following tips can help you tailor the project vision statement to meet your needs:

- Document the business goals that the product is aligned to.
- Identify key competitors and how this product will be better.
- Describe what differentiates this product from similar products in the market.
- You can combine all the information into a sentence or two, or you can follow a formula like the one shown in the sample template.

For hybrid projects, the vision statement can be combined with the project charter (see Section 1.5).

Alignment

The product vision should be aligned and consistent with the following documents:

- Business proposal
- Project startup canvas
- Project charter
- Project brief

Description

You can use the descriptions in Table 1.4 to assist you in developing the vision statement.

TABLE 1.4   Elements of a Project Vision Statement

Document Element	Description
Product or service	Identify the product or service being developed.
Target customer	The person or group who will buy or use the product.
Needs	The needs or requirements that the product will address.
Key attributes	A brief description of important features or functions.
Key benefit	Describe why the customer would buy the product or service.

VISION STATEMENT

Project Title: **Date Prepared:**

We are developing _____ for _____.

To respond to the following need(s):

-
-
-

This product responds to those needs by providing the following key attributes:

-
-
-
-

Customers will buy this product because of these benefits:

-
-
-
-

1.5 PROJECT CHARTER

The project charter is a document that formally authorizes a project. The project charter defines the reason for the project and assigns a project manager and his or her authority level for the project. The contents of the charter describe the project in high-level terms, such as

- Project purpose
- High-level project description
- Project boundaries
- Key deliverables
- High-level requirements
- Overall project risk
- Project objectives and related success criteria
- Summary milestone schedule
- Project budget
- Key stakeholders
- Project exit criteria
- Assigned project manager, responsibility, and authority level
- Project sponsor or other person(s) authorizing the project

The project charter can receive information from

- Project proposal
- Project startup canvas
- Business case

It provides information to

- Stakeholder register
- All elements of the project management plan
- Requirements documentation
- Requirements traceability matrix
- Project scope statement
- Stakeholder engagement plan

The project charter is developed once and is not usually changed unless there is a significant change in the environment, scope, schedule, resources, budget, or stakeholders.

Tailoring Tips

Consider the following tips to help you tailor the project charter to meet your needs:

- Combine the project charter with the project scope statement, especially if your project is small.
- If you are doing the project under contract, you can use the statement of work as the project charter in some cases.

For hybrid projects, you can combine the project vision statement and the project charter.

Alignment

The project charter should be aligned and consistent with the following documents:

- Business case
- Project scope statement
- Milestone schedule
- Budget
- Stakeholder register
- Risk register

Description

You can use the element descriptions in Table 1.5 to assist you in developing a project charter.

TABLE 1.5   Elements of a Project Charter

Document Element	Description
Project purpose	The reason the project is being undertaken. May refer to a business case, the organization's strategic plan, external factors, a contract agreement, or any other reason for performing the project.
High-level project description	A summary-level description of the project.
Project boundaries	Limits to the project scope. May include scope exclusions, or other limitations.
Key deliverables	The high-level project and product deliverables. These will be further elaborated in the project scope statement.
High-level requirements	The high-level conditions or capabilities that must be met to satisfy the purpose of the project. Describe the product features and functions that must be present to meet stakeholders' needs and expectations. These will be further elaborated in the requirements documentation.
Overall project risk	An assessment of the overall riskiness of the project. Overall risk can include the underlying political, social, economic, and technological volatility, uncertainty, complexity, and ambiguity. It pertains to the stakeholder exposure to variations in the project outcome.
Project objectives and related success criteria	Project objectives are usually established for at least scope, schedule, and cost. The success criteria identify the metrics or measurements that will be used to measure success.
	There may be additional objectives as well. Some organizations include quality, safety, and stakeholder satisfaction objectives.
Summary milestone schedule	Significant events in the project. Examples include the completion of key deliverables, the beginning or completion of a project phase, or product acceptance.
Preapproved financial resources	The amount of funding available for the project. May include sources of funding and annual funding limits.
Key stakeholder list	An initial, high-level list of people or groups that have influenced or can influence project success, as well as those who are influenced by its success. This can be further elaborated in the stakeholder register.
Project exit criteria	The performance, metrics, conditions, or other measurements that must be met to conclude the project.

Document Element	Description
Assigned project manager, responsibility, and authority level	The authority of the project manager with regard to staffing, budget management and variance, technical decisions, and conflict resolution.
	Examples of staffing authority include the power to hire, fire, discipline, accept, or not accept project staff.
	Budget management refers to the authority of the project manager to commit, manage, and control project funds. Variance refers to the variance level that requires escalation.
	Technical decisions describe the authority of the project manager to make technical decisions about deliverables or the project approach.
	Conflict resolution defines the degree to which the project manager can resolve conflict within the team, within the organization, and with external stakeholders.
Name and authority of the sponsor or other person(s) authorizing the project charter	The name, position, and authority of the person who oversees the project manager for the purposes of the project. Common types of authority include the ability to approve changes, determine acceptable variance limits, resolve inter-project conflicts, and champion the project at a senior management level.

PROJECT CHARTER

Project Title: _____ Date Prepared: _____

Project Sponsor: _____ Project Manager: _____

Project Purpose

High-Level Project Description

Project Boundaries

Key Deliverables

High-Level Requirements

Overall Project Risk

PROJECT CHARTER

	Project Objectives	Success Criteria
Scope		
Time		
Cost		
Other		

Summary Milestones	Due Date

Budget

Stakeholder(s)	Role

PROJECT CHARTER

Project Exit Criteria

Project Manager Authority Level

Staffing Decisions
Budget Management and Variance
Technical Decisions
Conflict Resolution

Approvals:

Project Manager Signature	Sponsor or Originator Signature
Project Manager Name	Sponsor or Originator Name
Date	Date

1.6 PROJECT BRIEF

The project brief is a short description that summarizes the key elements of the project. Ideally, it is one or two pages. A project brief is used to ensure that the project team and key stakeholders have a common understanding of the project.

Typical information includes

- Project overview
- Goals and objectives
- Success criteria
- Initial constraints and assumptions
- Scope description
- Budget
- Timeline and milestones
- Key stakeholders

The project brief can receive information from

- Project proposal
- Business case
- Project startup canvas
- Vision statement

It may provide information on

- Project management plan
- Scope statement
- Assumption log
- Project schedule
- Cost estimates
- Stakeholder engagement plan
- Stakeholder register

It is developed once, and then only changed if there are significant changes to scope, schedule, cost, or stakeholders.

Tailoring Tips

Consider the following tips to help tailor the project brief to meet your needs:

- If the project is being done for an external customer, include information on the customer and the target audience for the end product or service.
- For larger projects, the constraints and assumptions may be kept on an assumption log.

For hybrid projects, you may include information on methodologies that will be used and which deliverables will use adaptive methods and which will use predictive methods.

Alignment

The project brief should be aligned and consistent with the following documents:

- Project proposal
- Project charter
- Project roadmap
- Project management plan

Description

You can use the element descriptions in Table 1.6 to assist you in developing a project brief.

TABLE 1.6   Elements of a Project Brief

Document Element	Description
Project overview	A summary description of the project. This may include a short background on why the project is necessary.
Goals and objectives	The specific and measurable goals and objectives the project intends to achieve.
Success criteria	The measures or criteria that must be met for the goals and objectives for the project to be considered a success.
Initial constraints and assumptions	Project limitations (constraints) and expectations about the project.
Scope description	A short narrative of the scope.
Budget	The funds allocated to the project.
Timeline and milestones	The project duration and key milestone dates.
Key stakeholders	Significant stakeholders, including the customer and end user.

PROJECT BRIEF

Project Title: _____ Date: _____

Project Overview

Goals

Success Criteria

Objectives

Success Criteria

Constraints

Assumptions

Scope Description

PROJECT BRIEF

Budget:

Timeline:

Milestones

Key Stakeholders

1.7 PROJECT ROADMAP

The project roadmap is a high-level visual summary of the life cycle phases, key deliverables, management reviews, and milestones. Typical information includes

- Project life cycle phases
- Major deliverables or events in each phase
- Significant milestones
- Timing and types of reviews

The project roadmap can receive information from

- Project charter
- Project management plan

It provides information on

- Project schedule
- Risk register
- Milestone list

It is developed once, and then only changed if dates of the key events, milestones, or deliverables change.

Tailoring Tips

Consider the following tips to help tailor the project roadmap to meet your needs:

- For large and complex projects, this will likely be a separate stand-alone document.
- For smaller projects, the project roadmap may serve as the project management plan.
- For hybrid projects, you may include information on the development approach and release dates.

Alignment

The project roadmap should be aligned and consistent with the following documents:

- Project charter
- Project management plan

Description

You can use the element descriptions in Table 1.7 to assist you in developing a project roadmap.

TABLE 1.7   Elements of a Project Roadmap

Document Element	Description
Project life cycle phases	The name of each life cycle phase.
Major deliverables or events	Key deliverables, phase gates, key approvals, external events, or other significant events in the project.
Significant milestones	Milestones in the project.
Timing and types of reviews	Management, customer, compliance, or other significant reviews.

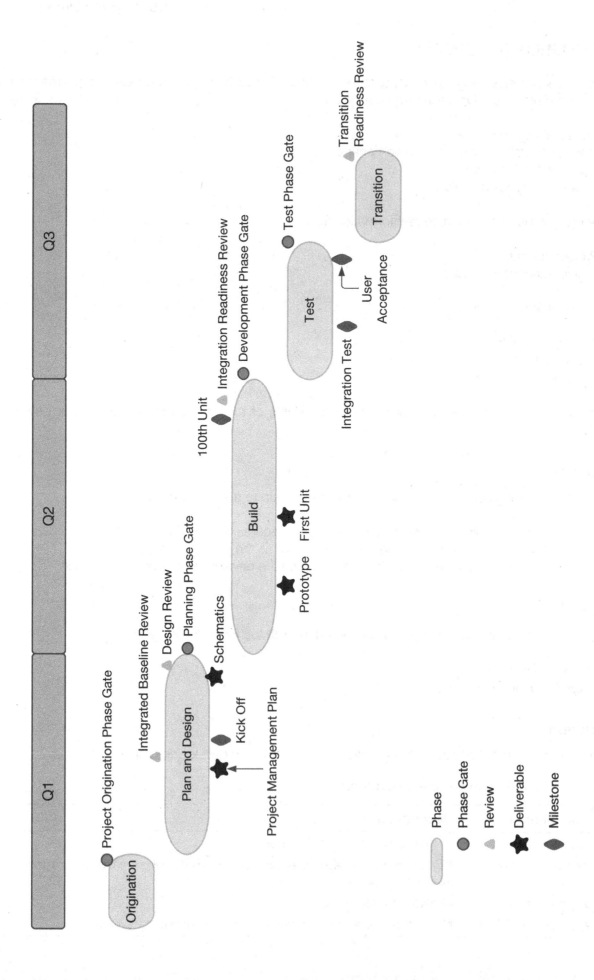

Project Plans

Once a project is authorized, the planning begins. Some projects with a well-defined scope that is unlikely to change spend a great deal of time planning up front and seek to minimize changes to the plans. Other projects do moderate planning at the start and then progressively elaborate the plans as more information becomes available. Projects that use Agile methods expect to evolve the scope and do just enough planning to start working and then evolve the scope of the project based on feedback and work priorities.

The management plan templates in this section are used predominately for projects with a well-defined scope that is not expected to change. The templates can be used to determine and document how various aspects of the project will be performed, such as communication management, risk management, and procurement management. There are 13 templates associated with planning the project:

- Scope management plan
- Requirements management plan
- Schedule management plan
- Release plan
- Cost management plan
- Quality management plan
- Resource management plan
- Communications plan
- Risk management plan
- Procurement management plan
- Stakeholder engagement plan
- Change management plan
- Project management plan

Some templates in this section can be combined to meet the needs of the project, stakeholders, and organization, such as the scope management plan and the quality management plan, or the communication plan and the stakeholder engagement plan.

For small projects, you may not need all these plans; for example, if there are no procurements, you would not need a procurement management plan. A small team with few physical resources may not need a resource management plan.

Project plans are usually developed once prior to baselining the project. If there is a significant change in the environment or the project, project plans may need to revisited.

2.1 SCOPE MANAGEMENT PLAN

The scope management plan is part of the project management plan. It specifies how the project scope will be defined, organized, developed, tracked, and validated. Planning how to manage scope should, at least, include processes for

- Decomposing the project into discrete deliverables
- Organizing the scope
- Determining what constitutes a scope change versus a revision
- How scope changes will be managed through the formal change control process
- Maintaining the scope baseline
- How deliverables will be accepted

In addition, the scope management plan may provide direction on the elements that should be contained in a work breakdown structure (WBS) dictionary and how the scope and requirements management plans interact.

The scope management plan can receive information from

- Project proposal
- Project charter
- Project brief
- Project management plan

It provides information to

- Requirements documentation
- Scope statement
- WBS
- WBS dictionary

It is developed once and does not usually change.

Tailoring Tips

Consider the following tips to help tailor the scope management plan to meet your needs:

- For smaller projects, you can combine the scope management plan with the requirements management plan.
- For larger projects, consider a test and evaluation plan that defines how deliverables will be validated and accepted by the customer.
- If your project involves business analysis, you may want to incorporate information on how business analysis activities and project management activities will interact.
- If you are using a hybrid approach, you can tailor the scope management plan by having two sections, one for predictive deliverables and one for adaptive deliverables. For the section on adaptive deliverables, you can include the following information:
  - Document how predictive and adaptive work will integrate
  - Identify how the backlog will be prioritized and maintained
  - Describe the format for user stories

Alignment

The scope management plan should be aligned and consistent with the following documents:

- Development approach
- Life cycle description
- Change management plan
- Requirements management plan
- Backlog
- User stories

Description

You can use the descriptions in Table 2.1 to assist you in developing a scope management plan.

TABLE 2.1   Elements of the Scope Management Plan

Document Element	Description
Decomposing and organizing scope	If a WBS will be used to decompose and organize the scope, describe whether it will be arranged using phases, geography, major deliverables, or some other way. The guidelines for establishing control accounts and work packages can also be documented in this section.
WBS dictionary	If a WBS dictionary is used, identify the information that will be documented and the level of detail required.
Scope change	Describe what constitutes a scope change versus a scope revision. A change must follow the change control process, whereas a revision does not.
Scope baseline maintenance	Identify the types of scope changes that will need to go through the formal change control process and how the scope baseline will be maintained.
Deliverable acceptance	For each deliverable, identify how the deliverable will be validated for customer acceptance, including any tests or documentation needed for sign-off.
Scope and requirements integration	Describe how project and product requirements will be addressed in the scope statement and WBS. Identify the integration points and how requirements and scope validation will occur.
Hybrid considerations	Describe how adaptive deliverables will be integrated with predictive deliverables.
Adaptive scope	Identify the type of backlog that will be used, e.g., requirements backlog, user story backlog, etc. Describe the criteria for prioritizing work in the backlog.
User stories	If working with user stories, identify the format the team should use.

SCOPE MANAGEMENT PLAN

Project Title: _____ Date: _____

Work Breakdown (WBS) Structure

WBS Dictionary

Scope Baseline Maintenance

Deliverable Acceptance

SCOPE MANAGEMENT PLAN

Scope and Requirements Integration

Hybrid Considerations

Adaptive Scope

User Stories

2.2 REQUIREMENTS MANAGEMENT PLAN

The requirements management plan is part of the project management plan. It specifies how requirements activities will be conducted throughout the project. Managing requirements activities includes, at least,

- Planning activities such as
  - Eliciting
  - Analyzing
  - Categorizing
  - Prioritizing
  - Documenting
  - Determining metrics
  - Defining the traceability structure

- Managing activities such as
  - Tracking
  - Reporting
  - Tracing
  - Validating
  - Performing configuration management

The requirements management plan can receive information from

- Project proposal
- Project charter
- Development approach
- Scope management plan
- Quality management plan

It provides information to

- Requirements documentation
- Requirements traceability matrix
- Quality management plan
- Risk register

The requirements management plan is developed once and does not usually change.

Tailoring Tips

Consider the following tips to help tailor the management plan to meet your needs:

- For smaller projects, you can combine the requirements management plan with the scope management plan.
- If your project involves business analysis, you may want to incorporate information on how business analysis requirements activities and project management requirements activities will interact.
- You can document the test and evaluation strategy in this plan. For larger projects, you may want to have a separate testing plan.

- For hybrid projects, you can specify how you will handle requirements for the predictive aspects and the adaptive aspects.
- If you are using an agile or adaptive development approach, you can incorporate information on how the backlog will be used to manage and track requirements.

Alignment

The requirements management plan should be aligned and consistent with the following documents:

- Development approach
- Change management plan
- Scope management plan
- Backlog

Description

You can use the descriptions in Table 2.2 to assist you in developing a requirements management plan.

TABLE 2.2   Elements of the Requirements Management Plan

Document Element	Description
Requirements collection	Describe how requirements will be collected or elicited. Consider techniques such as brainstorming, interviewing, observation, etc.
Requirements analysis	Describe how requirements will be analyzed for prioritization, categorization, and impact to the product or project approach.
Requirements categories	Identify categories for requirements such as business, stakeholder, quality, etc.
Requirements documentation	Define how requirements will be documented. The format of a requirements document may range from a simple spreadsheet to more elaborate forms containing detailed descriptions and attachments.
Requirements prioritization	Identify the prioritization approach for requirements. Certain requirements will be non-negotiable, such as those that are regulatory or those that are needed to comply with the organization's policies or infrastructure. Other requirements may be nice to have, but not necessary for functionality.
Requirements metrics	Document the metrics that requirements will be measured against. For example, if the requirement is that the product must be able to support 150 lb, the metric may be that it is designed to support 120 percent (180 lb) and that any design or engineering decisions that cause the product to go below the 120 percent need approval by the customer.
Requirements traceability	Identify the information that will be used to link requirements from their origin to the deliverables that satisfy them.
Requirements tracking	Describe how often and what techniques will be used to track progress on requirements.
Requirements reporting	Describe how reporting on requirements will be conducted and the frequency of such reporting.
Requirements validation	Identify the various methods that will be used to validate requirements such as inspection, audits, demonstration, testing, etc.
Requirements configuration management	Describe the configuration management system that will be used to control requirements, documentation, the change management process, and the authorization levels needed to approve changes.

REQUIREMENTS MANAGEMENT PLAN

Project Title: _____ **Date:** _____

Collection

Analysis

Categories

Documentation

Prioritization

REQUIREMENTS MANAGEMENT PLAN

Metrics

Traceability Structure

Tracking

Reporting

Validation

Configuration Management

2.3 SCHEDULE MANAGEMENT PLAN

The schedule management plan is part of the project management plan. It specifies how the project schedule will be developed, monitored, and controlled. Planning how to manage the schedule can include at the least

- Scheduling methodology
- Scheduling tool
- Level of accuracy for duration estimates
- Units of measure
- Variance thresholds
- Schedule updates

The schedule management plan can receive information from

- Project charter
- Project brief
- Project management plan

It provides information to

- Activity duration estimates
- Project schedule
- Schedule baseline
- Risk register

The schedule management plan is developed once and does not usually change.

Tailoring Tips

Consider the following tips to help tailor the schedule management plan to meet your needs:

- Add information on the level of detail and timing for WBS decomposition based on rolling wave planning.
- For projects that use earned value management, include information on rules for establishing percent complete and the earned value management (EVM) measurement techniques (fixed formula, percent complete, level or effort, etc.).
- For hybrid projects add information for iteration duration, iteration planning, and release planning.

Alignment

The schedule management plan should be aligned and consistent with the following documents:

- Project charter
- Cost management plan

Description

You can use the descriptions in Table 2.3 to assist you in developing a schedule management plan.

TABLE 2.3   Elements of the Schedule Management Plan

Document Element	Description
Schedule methodology	Identify the scheduling methodology that will be used for the project, whether it is critical path, Agile, or a combination of the two.
Scheduling tool(s)	Identify the scheduling tool(s) that will be used for the project. Tools can include scheduling software, reporting software, earned value software, etc.
Level of accuracy	Describe the level of accuracy needed for estimates. The level of accuracy may evolve over time as more information is known (progressive elaboration). If there are guidelines for rolling wave planning and the level of refinement that will be used for duration and effort estimates, indicate the levels of accuracy required as time progresses.
Units of measure	Indicate whether duration estimates will be in days, weeks, months, iterations, releases, or some other unit of measure.
Variance thresholds	Indicate the measures that determine whether an activity, work package, or the project as a whole is on time, requires preventive action, or is late and requires corrective action.
Schedule updates	Document the process for updating the schedule, including update frequency, permissions, and version control. Indicate the guidelines for maintaining baseline integrity and for re-baselining if necessary.
Iterations and releases	Document the duration for iterations. Describe how iteration planning and release planning will be conducted.

SCHEDULE MANAGEMENT PLAN

Project Title: _____ Date: _____

Schedule Methodology

Scheduling Tools

Level of Accuracy	Units of Measure	Variance Thresholds

Schedule Updates

SCHEDULE MANAGEMENT PLAN

Iterations and Releases

Release 1 Goal	
Iteration 1 Goal	
Iteration 2 Goal	
Iteration 3 Goal	
Release 2 Goal	
Iteration 1 Goal	
Iteration 2 Goal	
Iteration 3 Goal	

2.4 RELEASE PLAN

The release plan is similar to a roadmap. It functions as a high-level schedule that indicates which release each requirement or user story will be assigned to. The elements in a particular release can be updated based on the relative priority of the requirements in the backlog and the availability of resources needed to work on the specific requirements.

The release plan includes at the least

- Release goal
- Release dates
- User stories or requirements from the backlog

The release plan can be further elaborated into iterations. Each release has multiple iterations. Once an iteration starts, the requirements or user stories in the iteration cannot be changed.

The release plan can receive information from

- Project startup canvas
- Project charter
- Project brief
- Project roadmap
- Project management plan

It provides information to

- User stories
- Backlog

A high-level release plan may be developed after the product vision and backlog are started; however, it will remain somewhat dynamic throughout the project as priorities shift and new requirements are identified. The sequence and priorities of requirements or user stories may be updated after each iteration to reflect changing needs based on performance feedback or customer needs.

Tailoring Tips

Consider the following tips to help tailor the release plan to meet your needs:

- As information in the release plan gets more concrete, you may want to assign requirements or user stories to specific iterations in the release.
- You can arrange the release plan in "swim lanes," with each lane assigned to a specific team or work stream.
- If you show the release plan with a timeline, you can show the relationships between various user stories or features. This blends the information about the content of each release with the schedule information.
- For hybrid projects, you can incorporate releases as milestones in an integrated master schedule. Iterations can be entered into a predictive schedule without the detail of the user stories. This allows the schedule to show the predictive work and the adaptive work on the same schedule, and the dependencies between predictive and adaptive work, while still keeping the flexibility inherent in adaptive work.

Alignment

The release plan should be aligned and consistent with the following documents:

- Product vision
- Roadmap
- Backlog
- Schedule management plan

Description

You can use the element descriptions in Table 2.4 to assist you in developing the release plan.

TABLE 2.4   Elements of a Release Plan

Document Element	Description
Release goal	The expected features or functions that will be part of the release.
Release dates	Either a timeline or a milestone indicator of when the release and the iterations within a release will start and finish.
User stories	The requirements or user stories from the backlog.

Project Name_____

This diagram assumes that different color notes indicate different categories of user stories.

2.5 COST MANAGEMENT PLAN

The cost management plan is a part of the project management plan. It specifies how the project costs will be estimated, structured, monitored, and controlled. The cost management plan can include the following information:

- Level of accuracy for cost estimates
- Units of measure
- Variance thresholds
- Rules for performance measurement
- Cost reporting information and format
- Process for estimating costs
- Process for developing a time-phased budget
- Process for monitoring and controlling costs

In addition, the cost management plan may include information on the level of authority associated with cost and budget allocation and commitment, funding limitations, and options and guidelines on how and when costs get recorded for the project.

The cost management plan can receive information from the

- Project charter
- Schedule management plan
- Risk management plan

It provides information to

- Cost estimates
- Risk register

The cost management plan is developed once and does not usually change.

Tailoring Tips

Consider the following tips to help tailor the cost management plan to meet your needs:

- On smaller projects, often the project manager does not manage the budget. In those cases, you would not need this template.
- Units of measure for each type of resource may be indicated in the cost management plan or the resource management plan.
- For projects that use EVM, include information on rules for establishing percent complete, the EVM measurement techniques (fixed formula, percent complete, level or effort, etc.). For those that don't, delete this field.

Alignment

The cost management plan should be aligned and consistent with the following documents:

- Project charter
- Schedule management plan

Description

You can use the descriptions in Table 2.5 to assist you in developing a cost management plan.

TABLE 2.5   Elements of a Cost Management Plan

Document Element	Description
Units of measure	Indicate how each type of resource will be measured. For example, labor units may be measured in staff hours, days, or weeks. Physical resources may be measured in gallons, meters, tons, or whatever is appropriate for the material. Some resources are based on a lump sum cost each time they are used.
Level of precision	Indicate whether cost estimates will be rounded to hundreds, thousands, or some other measurement.
Level of accuracy	Describe the level of accuracy needed for estimates. The level of accuracy may evolve over time as more information is known (progressive elaboration). If there are guidelines for rolling wave planning and the level of refinement that will be used for cost estimates, indicate the levels of accuracy required as time progresses.
Organizational procedure links	Cost estimating and reporting should follow the numbering structure of the WBS. It may also need to follow the organization's code of accounts or other accounting and reporting structures.
Control thresholds	Indicate the measures that determine whether an activity, work package, or the project as a whole is on budget, requires preventive action, or is over budget and requires corrective action. Usually indicated as a percent deviation from the baseline.
Rules of performance measurement	Identify the level in the WBS where progress and expenditures will be measured. For projects that use EVM, indicate whether costs will be reported at the work package or control account level. Describe the measurement method that will be used, such as weighted milestones, fixed-formula, percent complete, etc. Document the equations that will be used to forecast estimates to complete (ETC) and estimates at completion (EAC).
Additional details	Describe variables associated with strategic funding choices, such as make or buy, buy or lease, borrowing funds versus using in-house funding, etc.

COST MANAGEMENT PLAN

Project Title: _____ Date Prepared: _____

Units of Measure	Level of Precision	Level of Accuracy

Organizational Procedure Links

Control Thresholds

Rules of Performance Measurement

Additional Details

2.6 QUALITY MANAGEMENT PLAN

The quality management plan is a component of the project management plan. It describes how applicable policies, procedures, and guidelines will be implemented to achieve the quality objectives for the project. Information in the quality management plan can include:

- Quality standards that will be used on the project
- Quality objectives
- Quality roles and responsibilities
- Deliverables and processes subject to quality review
- Quality control and quality management activities for the project
- Quality procedures applicable for the project

The quality management plan can receive information from:

- Project charter
- Requirements management plan
- Risk management plan
- Stakeholder engagement plan
- Requirements documentation
- Requirements traceability matrix
- Risk register

It provides information to:

- Scope management plan
- Cost estimates
- Resource management plan
- Procurement documents (request for proposal [RFP], request for quotation [RFQ])

The quality management plan is developed once and is not usually changed.

Tailoring Tips

Consider the following tips to help tailor the quality management plan to meet your needs:

- On smaller projects, quality, requirements, and scope are often handled as a single aspect, whereas in larger projects they are separated out and may have distinct roles and responsibilities for each aspect.
- In many industries there are specific standards that must be adhered to. Your quality management plan may reference these by citing specific regulations, or they may be integrated into organizational policies and procedures.
- Quality management planning must be consistent with your organization's quality policies, processes, and procedures.

Alignment

The quality management plan should be aligned and consistent with the following documents:

- Project charter
- Scope management plan
- Requirements management plan
- Resource management plan
- Procurement documents (RFP, RFQ, etc.)

Description

You can use the element descriptions in Table 2.6 to assist you in developing a quality management plan.

TABLE 2.6   Elements of a Quality Management Plan

Document Element	Description
Quality standards	Quality standards are usually industry or product driven. They may be ISO standards, IEEE, or some other regulatory or industry body.
Quality objectives	Quality objectives are the measures that must be achieved by the project or product components to meet the stakeholder needs. Objectives are the target you want to achieve. You may have metrics or specifications that provide a quantifiable measurement of success.
Quality roles and responsibilities	Define the roles necessary to conduct quality activities on the project and the responsibilities associated with each.
Deliverables and processes subject to quality review	The key deliverables that have metrics or measures associated with quality objectives. The processes used in the project that require verification or validation that they are being performed correctly, or in accordance with quality requirements or objectives.
Quality management approach	The approach that will be used to manage the quality process. Includes the timing and content of project and product quality audits.
Quality control approach	The approach that will be used to measure the product and the project performance to ensure the product meets the quality objectives.
Applicable quality procedures	Procedures that will be used for the project, such as • Nonconformance and rework • Corrective actions • Quality audits • Continuous improvement

QUALITY MANAGEMENT PLAN

Project Title: _____ **Date Prepared:** _____

Quality Standards

| |
| |
| |

Quality Objectives

Deliverable	Metric or Specification	Measure

Quality Roles and Responsibilities

Role	Responsibilities

QUALITY MANAGEMENT PLAN

Deliverables and Processes Subject to Quality Review

Deliverables	Processes

Quality Management Approach

Quality Control Approach

Applicable Quality Procedures

2.7 RESOURCE MANAGEMENT PLAN

The resource management plan is a component of the project management plan. It provides guidance on how team members and physical resources should be allocated, managed, and released. Information in the resource management plan includes the following:

* Estimating methods used to identify the type, number, and skill level of team resources
* Information on how project team members will be acquired, on-boarded, and released
* Roles and responsibilities associated with the project
* Project organizational chart
* Training requirements
* Methods used to identify the type, amount, and grade of physical resources
* Information on how physical resources will be acquired
* Methods for managing physical resources, such as inventory, supply chain, and logistics

The resource management plan can receive information from

* Project proposal
* Project charter
* Quality management plan
* Scope baseline
* Project schedule
* Requirements documentation
* Risk register
* Stakeholder register

It provides information to

* Project budget
* Responsibility assignment matrix
* Communications management plan
* Risk register
* Procurement management plan

The resource management plan is developed once and usually does not change.

Tailoring Tips

Consider the following tips to help tailor the resource management plan to meet your needs:

* If you need to bring in outside contractors for the project, you will need to include information on how to on-board them to the project. You will also need to consider how to ensure they have all the information they need but no access to proprietary data. This may include a non-disclosure agreement or similar forms.
* For any team or physical resources that are acquired from outside the organization, you will need to work with procurement policies for the organization and the project.
* Projects with large amounts of inventory, supplies, or material should either reference organizational policies regarding managing physical resources or provide sufficient detail to ensure appropriate control.
* Projects that use Agile approaches will only use a resource management plan for physical resources. The team usually self-manages, so they may use a team operating agreement to document how they will work together.

Alignment

The resource management plan should be aligned and consistent with the following documents:

- WBS
- Requirements documentation
- Quality management plan
- Procurement management plan

Description

You can use the element descriptions in Table 2.7 to assist you in developing a resource management plan.

TABLE 2.7   Elements of a Resource Management Plan

Document Element	Description
Team member identification	Methods used to identify the skill sets needed and the level of skill needed. This includes techniques to estimate the number of resources needed, such as information from past projects, parametric estimates, or industry standards.
Team member acquisition	Document how staff will be brought on to the project. Describe any differences between internal team members and contract team members with regard to on-boarding procedures.
Team member management	Document how team members will be managed and eventually released from the team. Management methods may vary depending on the relative authority of the project manager and whether team members are internal to the organization or contract staff. Team member release should include methods for knowledge transfer.
Project organizational chart	Create a hierarchy chart to show the project reporting and organizational structure.
Roles and responsibilities	Provide information on the following: **Role.** Identify the role or job title and a brief description of the role. **Authority.** Define the decision-making, approval, and influence levels for each role. Examples include alternative selection, conflict management, prioritizing, rewarding and penalizing, etc. **Responsibility.** Define the activities that each role carries out, such as job duties, processes involved, and the hand-offs to other roles. **Qualifications.** Describe any prerequisites, experience, licenses, seniority levels, or other qualifications necessary to fulfill the role. **Competencies.** Describe specific role or job skills and capacities required to complete the work. May include details on languages, technology, or other information necessary to complete the roles successfully.
Training requirements	Describe any required training on equipment, technology, or company processes. Include information on how and when training will be accomplished.
Physical resource identification	Methods used to identify the materials, equipment, and supplies needed to complete the work. This includes units of measure and techniques to estimate the amount of resources needed, such as information from past projects, parametric estimates, or industry standards.
Physical resource acquisition	Document how equipment, materials, and supplies will be acquired. This can include buy, lease, rent, or pull from inventory. In the event resources are acquired, ensure alignment with procurement management processes.
Physical resource management	Document how materials, equipment, and supplies will be managed to ensure they are available when needed. This can include appropriate inventory, supply chain, and logistics information.

RESOURCE MANAGEMENT PLAN

Project Title: _____ **Date Prepared:** _____

Team Member Identification and Estimates

Role	Number	Skill Level
1.	1.	1.
2.	2.	2.
3.	3.	3.
4.	4.	4.
5.	5.	5.

Process for Staff Acquisition	Process for Staff Release

Roles, Responsibilities, and Authority

Role	Responsibility	Authority
1.	1.	1.
2.	2.	2.
3.	3.	3.
4.	4.	4.
5.	5.	5.

Project Organizational Structure

RESOURCE MANAGEMENT PLAN

Training Requirements

Physical Resource Identification and Estimates

Resource	Amount	Grade
1.	1.	1.
2.	2.	2.
3.	3.	3.
4.	4.	4.
5.	5.	5.

Resource Acquisition

Resource Management

2.8 COMMUNICATION PLAN

The communication management plan is a component of the project management plan. It describes how project communications will be planned, structured, implemented, and monitored for effectiveness. Typical information includes

- Stakeholder communication requirements
- Information
- Method or media
- Time frame and frequency
- Sender
- Glossary of common terminology

In addition, methods for addressing sensitive or proprietary information and for updating the communications management plan may be included.

The communications management plan can receive information from

- Project charter
- Project brief
- Requirements documentation
- Resource management plan
- Stakeholder register
- Stakeholder engagement plan

It provides information to

- Stakeholder register
- Stakeholder engagement plan

The communications management plan is updated periodically throughout the project as stakeholders are added and leave the project and as communication needs emerge and shift.

Tailoring Tips

Consider the following tips to help tailor the communication management plan to meet your needs:

- When multiple organizations are working on a project, there will be additional information needed, such as
  - The person responsible for authorizing release of internal or confidential information
  - How different communication hardware, software, and technologies will be addressed to ensure information gets to everyone regardless of their communication infrastructure
  - If you have a multinational team, your plan will need to account for the business language, currency unit of measure, translation, and other factors required to ensure effective communication across multiple countries and cultures.
- On a project that has a significant communication component, you will want to identify the resources allocated for communication activities, the time requirements, and the budget allocated.
- For projects with complex communication needs, include a flowchart of the sequence of communication events.

Alignment

The communications management plan should be aligned and consistent with the following documents:

- Project schedule
- Stakeholder register
- Stakeholder engagement plan

Description

You can use the element descriptions in Table 2.8 to assist you in developing the communications management plan.

TABLE 2.8   Elements of a Communications Management Plan

Document Element	Description
Stakeholder communication requirements	The people or the groups of people who need to receive project information and their specific requirements.
Information	Describe the information to be communicated, including language, format, content, and level of detail.
Method or media	Describe how the information will be delivered; for example, email, meetings, web meetings, etc.
Time frame and frequency	List how often the information is to be provided and under what circumstances.
Sender	Insert the name of the person or the group that will provide the information.
Glossary of common terminology	List any terms or acronyms unique to the project or that are used in a unique way.

COMMUNICATIONS MANAGEMENT PLAN

Project Title: _____   **Date Prepared:** _____

Stakeholder	Information	Method	Timing or Frequency	Sender

Glossary of Terms or Acronyms:

Attach relevant communication diagrams or flowcharts.

Page 1 of 1

2.9 RISK MANAGEMENT PLAN

The risk management plan is a component of the project management plan. It describes how risk management activities will be structured and performed for both threats and opportunities. Typical information includes

- Methodology
- Roles and responsibilities for risk management
- Funding to identify, analyze, and respond to risk
- Frequency and timing for risk management activities
- Risk categories
- Stakeholder risk appetite
- Methods to track and audit risk management activities
- Definitions of probability
- Definitions of impact by objective
- Probability and impact matrix template

The risk management plan can receive information from

- Business case
- Project charter
- Project brief
- Project management plan (any component)
- Stakeholder register

It provides information to

- Cost management plan
- Quality management plan
- Risk register
- Stakeholder engagement plan

The risk management plan describes the approach for all risk management processes and provides key information needed to conduct those processes successfully.

The risk management plan is developed once and does not usually change.

Tailoring Tips

Consider the following tips to help tailor the risk management plan to meet your needs:

- For a small, simple, or short-term project, you can use a simplified risk register with a 3 × 3 probability and impact matrix. You would also include risk information in the project status report rather than a separate risk report.
- For larger, longer, and more complex projects, you will want to develop a robust risk management process, including a more granular probability and impact matrix, quantitative assessments for the schedule and budget baselines, risk audits, and risk reports.

- Projects that are using an Agile approach address risk at the start of each iteration and during the retrospective.
- Projects that are using an Agile approach may refer to risks as blockers or impediments.

Alignment

The risk management plan should be aligned and consistent with the following documents:

- Scope management plan
- Schedule management plan
- Cost management plan
- Quality management plan
- Resource management plan
- Procurement management plan
- Stakeholder engagement plan

Description

You can use the element descriptions in Table 2.9 to assist you in developing the risk management plan.

TABLE 2.9 Elements of a Risk Management Plan

Document Element	Description
Methodology	Describe the methodology or approach to risk management. This includes any tools, approaches, or data sources that will be used.
Roles and responsibilities	Document the roles and responsibilities for various risk management activities.
Risk categories	Identify categorization groups used to sort and organize risks. These can be used to sort risks on the risk register or for a risk breakdown structure, if one is used.
Risk management funding	Document the funding needed to perform the various risk management activities, such as utilizing expert advice or transferring risks to a third party. Also establishes protocols for establishing, measuring, and allocating contingency and management reserves.
Frequency and timing	Determine the frequency of conducting formal risk management activities and the timing of any specific activities.
Stakeholder risk appetite	Identify the risk thresholds of the organization(s) and key stakeholders on the project with regard to each objective.
Risk tracking and audit	Document how risk activities will be recorded and how risk management processes will be audited.
Definitions of probability	Document how probability will be measured and defined. Include the scale used and the definition for each level in the probability scale. The probability definitions should reflect the stakeholder risk appetite. For example: Very high = there is an 80% probability or higher that the event will occur High = there is a 60–80% probability that the event will occur Medium = there is a 40–60% probability that the event will occur Low = there is a 20–40% probability that the event will occur Very low = there is a 1–20% probability that the event will occur

(continued)

TABLE 2.9   Elements of a Risk Management Plan (*continued*)

Document Element	Description
Definitions of impact by objective	Document how impact will be measured and defined for either the project as a whole or for each objective. The probability definitions should reflect the stakeholder risk appetite. Include the scale used and the definition for each level in the impact scale. For example:
	Cost Impacts
	Very high = overrun of control account budget of >20%
	High = overrun of control account budget between 15–20%
	Medium = overrun of control account budget between 10–15%
	Low = overrun of control account budget between 5–10%
	Very low = overrun of control account budget of <5%
Probability and impact matrix	Describe the combinations of probability and impact that indicate a high risk, a medium risk, and a low risk and the scoring that will be used to prioritize risks. This can also include an assessment of proximity to indicate how soon the risk event is likely to occur.

RISK MANAGEMENT PLAN

Project Title: _____ **Date Prepared:** _____

Methodology

Roles and Responsibilities

Role	Responsibility
1.	1.
2.	2.
3.	3.
4.	4.

Risk Categories

Risk Management Funding

RISK MANAGEMENT PLAN

Contingency Protocols

Frequency and Timing

Stakeholder Risk Tolerances

Risk Tracking and Audit

RISK MANAGEMENT PLAN

Definitions of Probability

Very High	
High	
Medium	
Low	
Very Low	

Definitions of Impact by Objective

	Scope	Quality	Time	Cost
Very High				
High				
Medium				
Low				
Very Low				

Probability and Impact Matrix

Very High					
High					
Medium					
Low					
Very Low					
	Very Low	Low	Medium	High	Very High

2.10 PROCUREMENT MANAGEMENT PLAN

The procurement management plan is a component of the project management plan that describes the activities undertaken during the procurement process. It describes how all aspects of a procurement will be managed. Typical information includes the following:

- How procured work will be coordinated and integrated with other project work. Of specific interest is
  - Scope
  - Schedule
  - Documentation
  - Risk
- Timing of procurement activities
- Contract performance metrics
- Procurement roles, responsibility, and authority
- Procurement constraints and assumptions
- Legal jurisdiction and currency
- Risk management concerns, such as need for bonds or insurance
- Prequalified sellers lists

The procurement management plan can receive information from

- Project proposal
- Project charter
- Project startup canvas
- Stakeholder register
- Scope management plan
- Requirements documentation
- Requirements traceability matrix
- Scope baseline
- Project schedule
- Quality management plan
- Resource management plan
- Risk register

It provides information to

- Risk register
- Stakeholder register
- Procurement strategy

The procurement management plan is developed once and does not usually change.

Tailoring Tips

Consider the following tips to help tailor the procurement management plan to meet your needs:

- For a project that will be done using internal resources only, you do not need a procurement management plan.
- For a project where materials will be purchased and there is a standing purchase order with a vendor, you will not need a procurement management plan.

- For projects with a few procurements, consider combining this template with the procurement strategy.
- You may wish to combine the assumptions and constraints for procurements with the assumption log.
- Work with the contracting or legal department to ensure compliance with organizational purchasing policies.

Alignment

The procurement management plan should be aligned and consistent with the following documents:

- Scope management plan
- Requirements management plan
- Scope baseline
- Schedule management plan
- Project schedule
- Cost management plan
- Cost estimates
- Project budget
- Procurement strategy

Description

You can use the element descriptions in Table 2.10 to assist you in developing the procurement management plan.

TABLE 2.10 Elements of a Procurement Management Plan

Document Element	Description	
Procurement integration	Scope	Define how the contractor's WBS will integrate with the project WBS.
	Schedule	Define how the contractor's schedule will integrate with the project schedule, including milestones and long lead items.
	Documentation	Describe how contractor documentation will integrate with project documentation.
	Risk	Describe how risk identification, analysis, and response will integrate with risk management for the overall project.
	Reporting	Define how the contractor's status reports will integrate with the project status report.
Timing	Identify the timetable of key procurement activities. Examples include when the statement of work (SOW) will be complete, when procurement documents will be released, the date proposals are due, and so forth.	
Performance metrics	Document the metrics that will be used to evaluate the seller's performance.	
Roles, responsibilities, and authority	Define the roles, responsibilities, and authority level of the project manager, contractor, and procurement department, as well as any other significant stakeholders for the contract.	

(continued)

TABLE 2.10   Elements of a Procurement Management Plan (*continued*)

Document Element	Description
Assumptions and constraints	Record assumptions and constraints related to the procurement activities.
Legal jurisdiction and currency	Identify the location that has legal jurisdiction. Identify the currency that will be used for pricing and payment.
Risk management	Document requirements for performance bonds or insurance contracts to reduce risk.
Prequalified sellers	List any prequalified sellers that will be used.

PROCUREMENT MANAGEMENT PLAN

Project Title: _____ **Date:** _____

Procurement Integration

Area	Integration Approach
Scope	
Schedule	
Documentation	
Risk	
Reporting	

Timing of Key Procurement Activities

Date	Activity

Performance Metrics

Item	Metric	Measurement Method

PROCUREMENT MANAGEMENT PLAN

Roles, Responsibility, and Authority

Role	Responsibility	Authority

Assumptions and Constraints

Category	Assumption/Constraint

Legal Jurisdiction and Currency

Risk Management

Prequalified Sellers

1.
2.
3.
4.

2.11 STAKEHOLDER ENGAGEMENT PLAN

The stakeholder engagement plan is a component of the project management plan. It describes the strategies and actions that will be used to promote productive involvement of stakeholders in decision making and project performance. Typical information includes:

- Desired and current engagement level of key stakeholders
- Engagement approach for each stakeholder or group of stakeholders
- Scope and impact of change to stakeholders
- Interrelationships and potential overlap between stakeholders

The stakeholder engagement plan can receive information from

- Project charter
- Project brief
- Project vision statement
- Project startup canvas
- Stakeholder register
- Assumption log
- Change log
- Issue log
- Resource management plan
- Project schedule
- Communications management plan
- Risk management plan
- Risk register

It provides information to

- Requirements documentation
- Communication management plan
- Stakeholder register

The stakeholder engagement plan is updated periodically throughout the project as needed.

Tailoring Tips

Consider the following tips to help tailor the stakeholder engagement plan to meet your needs:

- For small projects, you may not need a stakeholder engagement plan. You can combine the information with the communication management plan as appropriate.
- Projects with multiple stakeholders with overlapping and intersecting relationships can benefit from having a stakeholder relationship map that shows the connections.
- For many high-risk projects, stakeholder engagement is critical to success. For projects with many stakeholders, complex interactions, and high-risk stakeholders, you will need to have a robust stakeholder engagement plan.
- For hybrid projects or projects that use Agile approaches, a lot of stakeholder engagement takes place in product demos. This information should be included in the stakeholder engagement plan.

Alignment

The stakeholder engagement plan should be aligned and consistent with the following documents:

- Stakeholder register
- Communications management plan
- Project schedule

Description

You can use the element descriptions in Table 2.11 to assist you in developing the stakeholder engagement plan.

TABLE 2.11   Elements of a Stakeholder Engagement Plan

Document Element	Description
Stakeholder engagement assessment matrix	Use information from the stakeholder register to document stakeholders. Document "current" stakeholder engagement level with a "C" and "desired" stakeholder engagement with a "D." A common format includes the following stakeholder participation descriptions: Unaware. Unaware of project and its potential impacts. Resistant. Aware of project and potential impacts and resistant to the change. Neutral. Aware of project yet neither supportive nor resistant. Supportive. Aware of project and potential impacts and supportive of change. Leading. Aware of project and potential impacts and actively engaged in ensuring project success.
Stakeholder engagement approach	Describe the approach you will use with each stakeholder to move them to the preferred level of engagement.
Stakeholder changes	Describe any pending additions, deletions, or changes to stakeholders and the potential impact to the project.
Interrelationships	List any relationships between and among stakeholder groups.

STAKEHOLDER ENGAGEMENT PLAN

Project Title: _____ **Date Prepared:** _____

Stakeholder	Unaware	Resistant	Neutral	Supportive	Leading

C = Current level of engagement   D = Desired level of engagement

Stakeholder Engagement Approach

Stakeholder	Approach

Pending Stakeholder Changes

Stakeholder Relationships

2.12 CHANGE MANAGEMENT PLAN

The change management plan is a component of the project management plan. It describes how change will be managed on the project. Typical information includes the following:

- Structure and membership of a change control board
- Definitions of change
- Change control board
  - Roles
  - Responsibilities
  - Authority
- Change management process
  - Change request submittal
  - Change request tracking
  - Change request review
  - Change request disposition

The change management plan is related to

- Change log
- Change request

It provides information to

- Project management plan

The change management plan is a part of the project management plan and is developed once and is not usually changed.

Tailoring Tips

Consider the following tips to help tailor the change management plan to meet your needs:

- If you only have a few product components or project documents that require configuration management, you may be able to combine change management and configuration management into one plan. Otherwise, you may want to have a separate configuration management plan that describes how you will name, track, and audit configurable items.
- The rigor and structure of your change management plan should reflect the product development approach. For predictive approaches, a rigorous change management approach is appropriate. For adaptive approaches, the change management plan should allow for evolving scope.
- For hybrid projects that have some deliverables that will use a predictive development approach, and some that will use an adaptive development approach, identify those deliverables that will be subject to change control.

Alignment

The change management plan should be aligned and consistent with the following documents:

- Project roadmap
- Scope management plan

- Requirements management plan
- Schedule management plan
- Cost management plan
- Quality management plan

Description

You can use the descriptions in Table 2.12 to assist you in developing a change management plan.

TABLE 2.12   Elements of a Change Management Plan

Document Element	Description	
Change management approach	Describe the degree of change control and how change control will integrate with other aspects of project management.	
Definitions of change	**Schedule Change:** Define a schedule change versus a schedule revision. Indicate when a schedule variance needs to go through the change control process to be re-baselined. **Budget Change:** Define a budget change versus a budget update. Indicate when a budget variance needs to go through the change control process to be re-baselined. **Scope Change:** Define a scope change versus progressive elaboration. Indicate when a scope variance needs to go through the change control process to be re-baselined. **Project Document Change:** Define when updates to project management documents or other project documents need to go through the change control process to be re-baselined.	
Change control board	Name	Individual's name
	Role	Position on the change control board
	Responsibility	Responsibilities and activities required of the role
	Authority	Authority level for approving or rejecting changes
Change control process	Change request submittal	Describe the process used to submit change requests, including who receives requests and any special forms, policies, or procedures that need to be used.
	Change request tracking	Describe the process for tracking change requests from submittal to final disposition.
	Change request review	Describe the process used to review change requests, including analysis of impact on project objectives such as schedule, scope, cost, etc.
	Change request outcome	Describe the possible outcomes, such as accept, defer, or reject.

CHANGE MANAGEMENT PLAN

Project Title: _____ **Date Prepared:** _____

Change Management Approach

| |
| |
| |

Definitions of Change

| Schedule change: |
| |
| Budget change: |
| |
| Scope change: |
| |
| Project document changes: |
| |

Change Control Board

Name	Role	Responsibility	Authority

CHANGE MANAGEMENT PLAN

Change Control Process

Change request submittal	
Change request tracking	
Change request review	
Change request disposition	

Attach relevant forms used in the change control process.

2.13 PROJECT MANAGEMENT PLAN

The project management plan describes how the team will perform, manage, and close the project. While it has some unique information, it is primarily comprised of all the subsidiary management plans and the baselines. The project management plan combines all this information into a cohesive and integrated approach to managing the project. Typical information includes

- Selected project life cycle
- Development approach for key deliverables
- Variance thresholds
- Baseline management
- Timing and types of reviews

The project management plan contains plans for managing all specific aspects of the project that require special focus. These take the form of subsidiary management plans and can include

- Change management plan
- Scope management plan
- Schedule management plan
- Requirements management plan
- Cost management plan
- Quality management plan
- Resource management plan
- Communication plan
- Risk management plan
- Procurement management plan
- Stakeholder engagement plan

The project management plan also contains baselines. Common baselines include:

- Scope baseline
- Schedule baseline
- Cost baseline
- Performance measurement baseline (an integrated baseline that includes scope, schedule, and cost)

In addition, any other relevant, project-specific information that will be used to manage the project is recorded in the project management plan.

The project management plan can receive information from all the startup documents, subsidiary management plans, and baselines. Because it is the foundational document for managing the project, it also provides information to all subsidiary plans.

The project management plan is first created as the initial project planning is conducted. Once it is baselined, it is not usually changed unless there is a significant change in the charter, environment, or scope of the project.

Tailoring Tips

Consider the following tips to help tailor the project management plan to meet your needs:

- For large and complex projects, each subsidiary management plan will likely be a separate stan-dalone plan. In this case, you may present your project management plan as a shell with just information on the life cycle, development approach, and key reviews, and then provide a link or reference to the more detailed subsidiary management plans.

- For smaller projects, a project roadmap that summarizes the project phases, major deliverables, milestones, and key reviews may be sufficient.
- You will likely have additional subsidiary management plans that are relevant to the nature of your project, such as a technology management plan, a logistics management plan, a safety management plan, and so forth.
- For projects that have an Agile component, the project management plan should describe how the deliverables that are using an Agile development approach are integrated with the parts of the project that are using a predictive development approach.

Alignment

The project management plan should be aligned and consistent with the following documents:

- Project charter
- Vision statement
- Project brief
- Project roadmap
- All subsidiary management plans

Description

You can use the element descriptions in Table 2.13 to assist you in developing a project management plan.

TABLE 2.13   Elements of a Project Management Plan

Document Element	Description
Project life cycle	Describe the life cycle that will be used to accomplish the project. This may include the following:
	• Name of each phase
	• Key activities for the phase
	• Key deliverables for the phase
	• Entry criteria for the phase
	• Exit criteria for the phase
	• Key reviews for the phase
Development approaches	Document the specific approach you will take to create key deliverables. Common approaches include predictive approaches, where the scope is known and stable, and adaptive approaches, where the scope is evolving and subject to change. It may also include iterative, incremental, or hybrid development approaches.
Subsidiary management plans	List the subsidiary management plans that are part of the project management plan. This can be in the form of a "table of contents," links to electronic copies of the subsidiary plans, or a list of the other plans that should be considered part of the project management plan but are separate documents.
Scope variance threshold	Define acceptable scope variances, variances that indicate a warning, and variances that are unacceptable. Scope variance can be indicated by the features and functions that are present in the end product, or the performance metrics that are desired.

(continued)

TABLE 2.13   Elements of a Project Management Plan (*continued*)

Document Element	Description
Scope baseline management	Describe how the scope baseline will be managed, including responses to acceptable, warning, and unacceptable variances. Define circumstances that would trigger preventive or corrective action and when the change control process would be enacted. Define the difference between a scope revision and a scope change. Generally, a revision does not require the same degree of approval that a change does. For example, changing the color of something is a revision; changing a function is a change.
Schedule variance threshold	Define acceptable schedule variances, variances that indicate a warning, and variances that are unacceptable. Schedule variances may indicate the percent of variance from the baseline or they may include the amount of float used or whether any schedule reserve has been used.
Schedule baseline management	Describe how the schedule baseline will be managed, including responses to acceptable, warning, and unacceptable variances. Define circumstances that would trigger preventive or corrective action and when the change control process would be enacted.
Cost variance threshold	Define acceptable cost variances, variances that indicate a warning, and variances that are unacceptable. Cost variances may indicate the percent of variance from the baseline, such as 0–5%, 5–10%, and greater than 10%.
Cost baseline management	Describe how the cost baseline will be managed, including responses to acceptable, warning, and unacceptable variances. Define circumstances that would trigger preventive or corrective action and when the change control process would be enacted.
Baselines	Attach all project baselines.

PROJECT MANAGEMENT PLAN

Project Title: _____ **Date Prepared:** _____

Project Life Cycle

Phase	Entry Criteria	Exit Criteria
1.		
2.		
3.		
4.		
5.		

Key Activities	Key Deliverables	Reviews

Development Approaches:

Deliverable	Development Approach

PROJECT MANAGEMENT PLAN

Subsidiary Management Plans

Name	Comment
Scope	
Time	
Cost	
Quality	
Resource	
Communications	
Risk	
Procurement	
Stakeholder	
Other Plans	

PROJECT MANAGEMENT PLAN

Variance Thresholds

Scope Variance Threshold	Scope Baseline Management
Schedule Variance Threshold	**Schedule Baseline Management**
Cost Variance Threshold	**Cost Baseline Management**

Baselines

Attach all project baselines.

Project Documents

There are a wide range of project documents. Many of them are used to support project planning, such as the scope statement, cost estimating worksheet, and stakeholder analysis. Other project documents are used as needed during the project, for example, a change request, procurement strategy, and source selection criteria.

Templates in this section are helpful for projects that use predictive approaches (e.g., waterfall methodology), as well as adaptive methods such as iterative and incremental development. There are 16 templates for project documents:

- Change request
- Requirements documentation
- Requirements traceability matrix
- Scope statement
- Work breakdown structure (WBS) dictionary
- Effort-duration estimates
- Effort-duration worksheets
- Cost estimates
- Cost estimating worksheet
- Responsibility assignment matrix
- Team charter
- Probability and impact assessment
- Risk data sheet
- Procurement strategy
- Source selection criteria
- Stakeholder analysis

Some of the templates in this section can be combined with other templates in this book. For example, in some projects it makes sense to combine the scope statement with the project charter, or the responsibility assignment matrix with the resource management plan. Other documents may not be used at all. Smaller projects won't need a WBS dictionary, and projects without procurements won't need a procurement strategy or source selection criteria.

Some project documents are unlikely to change, such as the responsibility assignment matrix; others are dynamic in nature, such as the risk data sheet, which is updated as more information about a risk is uncovered.

3.1 CHANGE REQUEST

A change request is used to change any aspect of the project. It can pertain to product components, documents, cost, schedule, or any other aspect of the project. Typical information includes

- Requestor
- Category
- Description of the proposed change
- Justification
- Impacts of the proposed change on

  o Scope
  o Quality
  o Requirements
  o Cost
  o Schedule
  o Project documents

- Comments

A change request can come from almost anyone. Upon completion, it is submitted to the change control board for review.

Tailoring Tips

Consider the following tips to help tailor the change request to meet your needs:

- For smaller projects, you can simplify the template by having a summary description of the impacts without including impacts for each subcategory (scope, quality, requirements, etc.).
- You can add a check box that indicates whether the change is mandatory (such as a legal requirement) or discretionary.
- A field can be added that describes the implications of not making the change.

Alignment

The change request should be aligned and consistent with the following documents:

- Change management plan
- Change log

Description

You can use the element descriptions in Table 3.1 to assist you in developing the change request.

TABLE 3.1   Elements of a Change Request

Document Element	Description		
Requestor	The name, and if appropriate, the position of the person requesting the change		
Category	Check a box to indicate the category of change.		
Description of change	Describe the proposed change in enough detail to clearly communicate all aspects of the change.		
Justification for proposed change	Indicate the reason for the change.		
Impacts of change	Scope	Describe the impact of the proposed change on the project and product scope.	
	Quality	Describe the impact of the proposed change on the project or product quality.	
	Requirements	Describe the impact of the proposed change on the project or product requirements.	
	Cost	Describe the impact of the proposed change on the project budget, cost estimates, or funding requirements.	
	Schedule	Describe the impact of the proposed change on the schedule and whether it will change the critical path.	
Comments	Provide any comments that will clarify information about the requested change.		

CHANGE REQUEST

Project Title: _____ Date Prepared: _____

Requestor _____

Category _____

☐ Scope   ☐ Quality   ☐ Requirements
☐ Cost   ☐ Schedule   ☐ Documents

Detailed Description of Proposed Change

Justification for Proposed Change

Impacts of Change

Scope	☐ Increase	☐ Decrease	☐ Modify
Description:			

Quality	☐ Increase	☐ Decrease	☐ Modify
Description:			

CHANGE REQUEST

Requirements	☐ Increase	☐ Decrease	☐ Modify

Description:

Cost	☐ Increase	☐ Decrease	☐ Modify

Description:

Schedule	☐ Increase	☐ Decrease	☐ Modify

Description:

Stakeholder Impact	☐ High risk	☐ Low risk	☐ Medium risk

Description:

Comments

Disposition	☐ Approve	☐ Defer	☐ Reject

Justification

3.2 REQUIREMENTS DOCUMENTATION

Project success is directly influenced by the discovery and decomposition of stakeholders' needs into requirements and by the care taken in determining, documenting, and managing the requirements of the product, service, or result of the project.

These requirements need to be documented in enough detail to be included in the scope baseline and be measured and validated. Requirements documentation assists the project manager in making tradeoff decisions among requirements and in managing stakeholder expectations. Requirements will be progressively elaborated as more information about the project becomes available.

When documenting requirements, it is useful to group them by category. Some common categories include

- Business requirements
- Stakeholder requirements
- Solution requirements
- Transition and readiness requirements
- Project requirements
- Quality requirements

Requirements documentation should include at least

- Identifier
- Requirement
- Stakeholder
- Category
- Priority
- Acceptance criteria
- Test or verification method
- Release or phase

Requirements documentation can receive information from

- Project proposal
- Project charter
- Project brief
- Assumption log
- Stakeholder register
- Scope management plan
- Requirements management plan

It provides information to

- Stakeholder register
- Scope baseline
- Quality management plan
- Resource management plan
- Communications management plan
- Risk register
- Procurement management plan
- Project closeout report

Requirements documentation begins at a high level and may be elaborated as the project progresses. For projects that have a well-defined scope, the high-level requirements are not likely to change, merely be elaborated. For projects that are adaptive, the requirements documentation can evolve and change throughout the project.

Tailoring Tips

Consider the following tips to help tailor the requirements documentation to meet your needs:

- If you are using an Agile or adaptive development approach, you may want to incorporate information on the release or iteration for each requirement.
- For a project with a lot of requirements, you may want to indicate the relationships between requirements using a requirements traceability matrix or other tool.
- For projects with many requirements, or complex requirements, it may be better to automate the requirements management process.
- You can add information about assumptions or constraints associated with requirements.
- For small and quick adaptive or Agile projects, the requirements documentation and backlog can be combined.

Alignment

The requirements documentation should be aligned and consistent with the following documents:

- Requirements management plan
- Quality management plan
- Requirements traceability matrix
- Release plan

Description

You can use the descriptions in Table 3.2 to assist you in developing requirements documentation.

TABLE 3.2   Elements of Requirements Documentation

Document Element	Description
ID	A unique identifier for the requirement
Requirement	The condition or capability that must be met by the project or be present in the product, service, or result to satisfy a need or expectation of a stakeholder
Stakeholder	Stakeholder's name. If you don't have a name you can substitute a position or organization until you have more information.
Category	The category of the requirement
Priority	The priority group, for example, Level 1, Level 2, etc., or must have, should have, or nice to have
Acceptance criteria	The criteria that must be met for the stakeholder to approve that the requirement has been fulfilled
Test or verification method	The means that will be used to verify that the requirement has been met. This can include inspection, test, demonstration, or analysis.
Phase or release	The phase or release in which the requirement will be met

REQUIREMENTS DOCUMENTATION

Project Title: _____ Date Prepared: _____

ID	Requirement	Stakeholder	Category	Priority	Acceptance Criteria	Test or Verification	Phase or Release

3.3 REQUIREMENTS TRACEABILITY MATRIX

A requirements traceability matrix is used to track the various attributes of requirements throughout the project life cycle. It uses information from the requirements documentation and traces how those requirements are addressed through other aspects of the project. The following template shows how requirements would be traced to project objectives, WBS deliverables, and how they will be validated.

Another way to use the matrix is to trace the relationship between categories of requirements. For example,

* Business objectives and technical requirements
* Functional requirements and technical requirements
* Requirements and verification method
* Technical requirements and WBS deliverables

An inter-requirements traceability matrix can be used to record this information. A sample template is included after the requirements traceability matrix.

The requirements traceability matrix can receive information from

* Project charter
* Assumption log
* Stakeholder register
* Scope management plan
* Requirements management plan
* Stakeholder engagement plan
* Lessons learned register

It provides information to

* Quality management plan
* Procurement statement of work
* Change requests

The requirements traceability matrix is usually created once, and then updated if and when requirements are added or modified.

Tailoring Tips

Consider the following tips to help tailor the requirements traceability matrix to meet your needs:

* For complex projects, you may need to invest in requirements management software to help manage and track requirements. Using a paper form is usually only helpful for small projects or when tracking requirements at a high level.
* For projects with one or more vendors, you may want to add a field indicating which organization is accountable for meeting each requirement.
* Consider an outline format with the business requirement at a parent level and technical requirement and specifications subordinate to the business requirement.
* For projects that use adaptive processes or releases, you can include a column that indicates which release the requirement will be part of.
* For projects that use a backlog, the priority indicator should align with the sequence of items on the backlog.

Alignment

The requirements traceability matrix should be aligned and consistent with the following documents:

- Development approach
- Requirements management plan
- Requirements documentation
- Release and iteration plan
- Requirements or user story backlog

Description

You can use the element descriptions in Tables 3.3A and 3.3B to assist you in developing a requirements traceability matrix and an inter-requirements traceability matrix. The matrix shown uses an example of business and technical requirements.

TABLE 3.3A   Requirements Traceability Matrix

Document Element	Description
ID	Enter a unique requirement identifier.
Requirement	Document the condition or capability that must be met by the project or be present in the product, service, or result to satisfy a need or expectation of a stakeholder.
Source	The stakeholder that identified the requirement.
Priority	Prioritize the requirement category, for example, Level 1, Level 2, etc., or must have, should have, or nice to have.
Category	Categorize the requirement. Categories can include functional, nonfunctional, maintainability, security, etc.
Objective	List the business objective as identified in the charter or business case that is met by fulfilling the requirement.
Deliverable	Identify the deliverable that is associated with the requirement.
Verification	Describe the metric that is used to measure the satisfaction of the requirement.
Validation	Describe the technique that will be used to validate that the requirement meets the stakeholder needs.

TABLE 3.3B   Inter-Requirements Traceability Matrix

Document Element	Description
ID	Enter a unique business requirement identifier.
Business requirement	Document the condition or capability that must be met by the project or be present in the product, service, or result to satisfy the business needs.
Priority	Prioritize the business requirement category, for example, Level 1, Level 2, etc., or must have, should have, or nice to have.
Source	Document the stakeholder who identified the business requirement.
ID	Enter a unique technical requirement identifier.

TABLE 3.3B   Inter-Requirements Traceability Matrix (*continued*)

Document Element	Description
Technical requirement	Document the technical performance that must be met by the deliverable to satisfy a need or expectation of a stakeholder.
Priority	Prioritize the technical requirement category, for example, Level 1, Level 2, etc., or must have, should have, or nice to have.
Source	Document the stakeholder who identified the technical requirement.

INTER-REQUIREMENTS TRACEABILITY MATRIX

Project Title: _____   **Date Prepared:** _____

ID	Business Requirement	Priority	Source	ID	Technical Requirement	Priority	Source

INTER-REQUIREMENTS TRACEABILITY MATRIX

Project Title: _____ **Date Prepared:** _____

ID	Business Requirement	Priority	Source	ID	Technical Requirement	Priority	Source

3.4 PROJECT SCOPE STATEMENT

The project scope statement assists in defining and developing the project and product scope. It is used to provide a narrative description of the project and each of the key deliverables. The description provides more context than the WBS and requirements documentation and more detail than the project charter. The project scope statement should contain at least this information:

- Project scope description
- Project deliverables
- Product acceptance criteria
- Project exclusions

The project scope statement can receive information from

- Project proposal
- Project charter
- Project brief
- Project startup canvas
- Scope management plan
- Requirements documentation

It provides information to

- WBS

The project scope statement is developed once and is not usually updated unless there is a significant change in scope.

Tailoring Tips

Consider the following tips to help tailor the project scope statement to meet your needs:

- For smaller projects, you can combine the project scope statement with the project charter.
- For Agile projects, you can combine the information with the release and iteration plan.

Alignment

The project scope statement should be aligned and consistent with the following documents:

- Project charter
- Project proposal
- Project charter
- Project brief
- Project startup canvas
- Work breakdown structure
- Requirements documentation

Description

You can use the element descriptions in Table 3.4 to assist you in developing a project scope statement.

TABLE 3.4   Elements of a Project Scope Statement

Document Element	Description
Project scope description	Project scope is progressively elaborated from the project description in the project charter and the requirements in the requirements documentation.
Project deliverables	Project deliverables are progressively elaborated from the project description in the project charter.
Product acceptance criteria	Acceptance criteria is progressively elaborated from the information in the project charter. Acceptance criteria can be developed for each component of the project.
Project exclusions	Project exclusions clearly define what is out of scope for the product and project.

SCOPE STATEMENT

Project Title: _____ Date Prepared: _____

Project Scope Description

Project Deliverables	Acceptance Criteria

Project Exclusions

3.5 WBS DICTIONARY

The WBS dictionary supports the work breakdown structure (WBS). The work breakdown structure (WBS) is used to decompose all the work of the project. It begins at the project level and is successively broken down into finer levels of detail. The WBS can be shown as a chart or an outline. The lowest level is called a work package. Above the work package level are control accounts that are used for reporting progress. Because the WBS depends on the nature of the work, there are no templates included in this book.

The WBS dictionary supports the WBS by providing details about the control accounts and work packages it contains. The dictionary can provide detailed information about each work package or summary information at the control account level. Information in the WBS dictionary can include

- Work package name
- Description of work
- Schedule milestones
- Associated schedule activities
- Resources required
- Cost estimates
- Quality requirements
- Acceptance criteria
- Technical information or references
- Contract information

The WBS dictionary is progressively elaborated as the planning processes progress. Once the WBS is developed, the statement of work for a particular work package may be defined, but the necessary activities, cost estimates, and resource requirements may not be known. Thus, the inputs for the WBS dictionary are more detailed than for the WBS.

Use the information from your project to tailor the template to best meet your needs.

The WBS dictionary can receive information from

- Requirements documentation
- Project scope statement
- WBS
- Cost estimates
- Quality metrics
- Contracts

The WBS dictionary provides information to

- Risk register
- Procurement management plan

The WBS dictionary is progressively elaborated throughout the project.

Tailoring Tips

Consider the following tips to help tailor the WBS dictionary to meet your needs:

- For smaller projects, you may not need a WBS dictionary.
- For projects that do use a WBS dictionary, you can tailor the information to be as detailed or as high level as you need. You may just want to list a description of work, the cost estimate, key delivery dates, and assigned resources.

- For projects that have deliverables outsourced, you can consider the WBS dictionary as a mini-statement of work for the outsourced deliverables.
- Projects that use WBS dictionaries can reference other documents and the relevant sections for technical, quality, or agreement information rather than recording the information in the dictionary.

Alignment

The WBS dictionary should be aligned and consistent with the following documents:

- WBS
- Project scope statement
- Duration estimates
- Project schedule
- Cost estimates
- Project budget
- Quality management plan
- Resource management plan
- Procurement management plan

Description

You can use the element descriptions in Table 3.5 to assist you in developing a WBS dictionary.

TABLE 3.5   Elements of a WBS Dictionary

Document Element	Description
Work package name	Enter the name of the work package from the WBS.
Description of work	Enter a brief description of the work package deliverable from the WBS.
Milestones	List any milestones associated with the work package.
Due dates	List the due dates for milestones associated with the work package.
ID	Enter a unique activity identifier—usually an extension of the WBS numbering structure.
Activity	Enter the activity from the schedule.
Team resource	Identify the team members who will be involved with the work package.
Labor hours	Enter the total effort required.
Labor rate	Enter the labor rate, usually from cost estimates.
Labor total	Multiply the effort hours times the labor rate.
Material units	Enter the amount of material required.
Material cost	Enter the material cost, usually from cost estimates.
Material total	Multiply the material units times the material cost.
Total cost	Sum the labor, materials, and any other costs associated with the work package.
Quality requirements	Document any quality requirements or metrics associated with the work package.
Acceptance criteria	Describe the acceptance criteria for the deliverable, usually from the scope statement.
Technical information	Describe or reference any technical requirements or documentation needed to complete the work package.
Contract information	Reference any contracts or other agreements that impact the work package.

WBS DICTIONARY

Project Title: _____ Date Prepared: _____

Work Package Name	Description of Work

Milestones	Due Dates
1.	
2.	
3.	

ID	Activity	Resource	Labor			Material			Total Cost
			Hours	Rate	Total	Units	Cost	Total	

Quality Requirements

Acceptance Criteria

Technical Information

Contract Information

3.6 EFFORT/DURATION ESTIMATES

Effort estimates reflect the amount of work it takes to complete an activity. Duration estimates reflect the amount of time it takes to complete an activity. For those activity durations driven by human resources, as opposed to material or equipment, the duration estimates will generally convert the estimate of effort hours into days or weeks. To convert effort hours into days, take the total number of hours and divide by 8. To convert to weeks, take the total number of hours and divide by 40.

You can develop effort and duration estimates for each activity, work package, or control account. Duration estimates include at least

- ID
- Activity description
- Duration estimate
- Effort hours (Optional)

Duration estimates can receive information from

- Assumption log
- Schedule management plan
- Lessons learned register

It provides information to

- Project schedule

Duration estimates are developed throughout the project as schedule and activity details are refined.

Tailoring Tips

Consider the following tips to help tailor the duration estimates to meet your needs:

- At a control account level or phase level, duration estimates may include contingency reserve to account for risks related to uncertainty in the duration estimates, ambiguity in the scope, or resource availability.
- Develop duration estimates at the level of accuracy that suits your project needs. Rolling wave planning is often used for duration estimating; as more information is known about the project activities, duration estimates are refined and updated.
- For projects that are using an agile development approach, set time boxes are used rather than duration estimates. Additionally, different estimating methods are used.

Alignment

The duration estimates should be aligned and consistent with the following documents:

- Assumption log
- Cost estimates

Description

You can use the descriptions in Table 3.6 to assist you in developing the duration estimates.

TABLE 3.6   Elements of Duration Estimates

Document Element	Description
ID	Unique identifier
Activity description	A description of the work that needs to be done
Effort hours	The amount of labor it will take to accomplish the work; usually shown in hours, but may be shown in days
Duration estimates	The length of time it will take to accomplish the work; usually shown in days, but may be shown in weeks or months

DURATION ESTIMATES

Project Title: _____ Date Prepared: _____

ID	Activity Description	Effort Hours	Duration Estimate

3.7 EFFORT—DURATION ESTIMATING WORKSHEET

An effort or duration estimating worksheet can help to develop estimates when quantitative methods are used. Quantitative methods include

- Parametric estimates
- Analogous estimates
- Multi-point estimates

Parametric estimates are derived by determining the effort hours needed to complete the work. The effort hours are then calculated by

- Dividing the estimated hours by resource quantity (i.e., number of people assigned to the task)
- Dividing the estimated hours by the percent of time the resource(s) are available (i.e., 100% of the time, 75% of the time, or 50% of the time)
- Multiplying the estimated hours by a performance factor. Experts in a field generally complete work faster than people with an average skill level or novices. Therefore, a factor to account for the productivity is developed.

These estimates can be made even more accurate by considering that most people are productive on project work only about 75% of the time.

Analogous estimates are derived by comparing current work to previous similar work. The size of the previous work and the duration are compared to the expected size of the current work compared to the previous work. Then the ratio of the size of the current work is multiplied by the previous duration to determine an estimate. Various factors, such as complexity, can be factored in to make the estimate more accurate. This type of estimate is generally used to get a high-level estimate when detailed information is not available.

A multi-point estimate can be used to account for uncertainty in the duration estimate. Stakeholders provide estimates for optimistic (O), most likely (M), and pessimistic (P) scenarios. These estimates are put into an equation to determine an expected duration. The needs of the project determine the appropriate equation, but a common equation is the beta distribution:

$$\text{Estimated duration} = \frac{O + 4M + P}{6}$$

In formulas, duration is often represented by "*t*" for "time."

The duration estimating worksheet can receive information from

- Assumption log
- Scope and requirements information
- Schedule management plan
- Risk register
- Lessons learned register

Duration estimates are developed throughout the project as schedule and activity details are refined.

Tailoring Tips

Consider the following tips to help tailor the duration estimates to meet your needs:

- You can tailor the multi-point estimate to meet the needs of your project. For example, you can weight the pessimistic option more heavily if the team has not done this type of work before, or if there are risks associated with the work. For example, you may use an equation like this one:

$$\frac{O + 4M + 3P}{8}$$

Description

You can use the element descriptions in Table 3.7 to assist you in estimating durations with the worksheet.

TABLE 3.7   Elements of an Activity Duration Estimating Worksheet

Document Element	Description
ID	Unique identifier
Parametric estimates	
Effort hours	Enter amount of labor it will take to accomplish the work. Usually shown in hours, but may also be shown in days. Example: 150 hours
Resource quantity	Document the number of resources available. Example: 2 people
Percent available	Enter amount of time the resources are available. Usually shown as the percent of time available per day or per week. Example: 75% of the time
Performance factor	Estimate a performance factor if appropriate. Generally, effort hours are estimated based on the amount of effort it would take the average resource to complete the work. This can be modified if you have a highly skilled resource or someone who has very little experience. The more skilled the resource, the lower the performance factor. For example, an average resource would have a 1.0 performance factor. A highly skilled resource could get the work done faster, so you multiply the effort hours times a performance factor of .8. A less-skilled resource will take longer to get the work done, so you would multiply the effort hours times 1.2.
Duration estimate	Divide the effort hours by the resource quantity times the percent available times the performance factor to determine the length of time it will take to accomplish the work. The equation is Effort/(quantity × percent available × performance factor) = duration 150 hours of effort with 2 people assigned, that are available 75% of the time and are very experienced would be calculated like this: 150/(2 × .75 × .8) = 125 hours of duration
Analogous estimates	
Previous activity	Enter a description of the previous activity. Example: Build a 160 square foot deck.
Previous duration	Document the duration of the previous activity. Example: 10 days

TABLE 3.7 Elements of an Activity Duration Estimating Worksheet (*continued*)

Document Element	Description
Current activity	Describe how the current activity is different. Example: Build a 200 square foot deck.
Multiplier	Divide the current activity by the previous activity to get a multiplier. Example: 200/160 = 1.25
Duration estimate	Multiply the duration for the previous activity by the multiplier to calculate the duration estimate for the current activity. Example: 10 days × 1.25 = 12.5 days
Multi-point estimate (Beta distribution)	
Optimistic duration	Determine an optimistic duration estimate. Optimistic estimates assume everything will go well and there won't be any delays in material, and that all resources are available and will perform as expected. Example: 20 days
Most likely duration	Determine a most likely duration estimate. Most likely estimates assume that there will be some delays but nothing out of the ordinary. Example: 25 days
Pessimistic duration	Determine a pessimistic duration estimate. Pessimistic estimates assume there are significant risks that will materialize and cause delays. Example: 36 days
Weighting equation	Weight the three estimates and divide. The most common method of weighting is the beta distribution: $tE = (tO + 4tM + tP)/6$ Example: (20 + 4(25) + 36)/6
Expected duration	Enter the expected duration based on the beta distribution calculation. Example: 26 days

DURATION ESTIMATING WORKSHEET

Project Title:_____ Date Prepared: _____

Parametric Estimates

ID	Effort Hours	Resource Quantity	% Available	Performance Factor	Duration Estimate

Analogous Estimates

ID	Previous Activity	Previous Duration	Current Activity	Multiplier	Duration Estimate

Multi-Point Estimates

ID	Optimistic Duration	Most Likely Duration	Pessimistic Duration	Weighting Equation	Expected Duration Estimate

3.8 COST ESTIMATES

Cost estimates provide information on the cost of resources necessary to complete project work, including labor, equipment, supplies, services, facilities, and material. Estimates can be determined by developing an approximation for each work package using expert judgment or by using a quantitative method, such as

- Parametric estimates
- Analogous estimates
- Multi-point estimates

Cost estimates should include at least

- ID
- Labor costs
- Physical resource costs
- Reserve
- Estimate
- Basis of estimates
- Method
- Assumptions
- Range
- Confidence level

Cost estimates can receive information from

- Cost management plan
- Quality management plan
- Risk register
- Lessons learned register

They provide information to

- Project budget

Cost estimates are developed and then refined periodically as needed.

Tailoring Tips

Consider the following tips to help tailor the cost estimates to meet your needs:

- Cost estimates may include contingency reserve to account for risks related to uncertainty in the estimates or ambiguity in the scope or resource availability.
- If considerations for the cost of quality, cost of financing, or indirect costs were included, add that information to your cost estimate.
- Estimate costs at the level of accuracy and precision that suits your project needs. Rolling wave planning is often used for cost estimating; as more information is known about the scope and resources, cost estimates are refined and updated.
- If using vendors, indicate the estimated cost and indicate the type of contract being used to account for possible fees and awards.

Alignment

The cost estimates should be aligned and consistent with the following documents:

- Assumption log
- Effort estimates

Description

You can use the descriptions in Table 3.8 to assist you in developing the cost estimates.

TABLE 3.8   Elements of Cost Estimate

Document Element	Description
ID	Unique identifier, such as the WBS ID or activity ID
Resource	The resource (person, equipment, material) needed for the WBS deliverable
Labor costs	The costs associated with team or outsourced resources
Physical costs	Costs associated with material, equipment, supplies, or other physical resources
Reserve	Document contingency reserve amounts, if any
Estimate	The sum of the cost of labor, physical resources, and reserve costs
Basis of estimates	Information such as cost per pound, duration of the work, square feet, etc.
Method	The method used to estimate the cost, such as analogous, parametric, etc.
Assumptions/constraints	Assumptions used to estimate the cost, such as the length of time the resource will be needed
Range	The range of estimate
Confidence level	The degree of confidence in the estimate

3.9 COST ESTIMATING WORKSHEET

A cost estimating worksheet can help to develop cost estimates when quantitative methods or a bottom-up estimate are developed. Quantitative methods include:

- Parametric estimates
- Analogous estimates
- Multi-point estimates

Parametric estimates are derived by determining the cost variable that will be used and the cost per unit. The number of units is multiplied by the cost per unit to derive a cost estimate.

Analogous estimates are derived by comparing current work to previous similar work. The size of the previous work and the cost are compared to the expected size of the current work. Then the ratio of the size of the current work compared to the previous work is multiplied by the previous cost to determine an estimate. Various factors, such as complexity and price increases, can be factored in to make the estimate more accurate. This type of estimate is generally used to get a high-level estimate when detailed information is not available.

A multi-point estimate can be used to account for uncertainty in the cost estimate.

Stakeholders provide estimates for optimistic (O), most likely (M), and pessimistic (P) scenarios. These estimates are put into an equation to determine an expected duration. The needs of the project determine the appropriate equation, but a common equation is the beta distribution:

$$\text{Estimated duration} = \frac{O + 4M + P}{6}$$

In formulas, duration is often represented by "c" for "cost."

Bottom-up estimates are detailed estimates done at the work package level. Detailed information on the work package, such as technical requirements, engineering drawings, labor duration, and other direct and indirect costs, are used to determine the most accurate estimate possible.

The cost estimating worksheet can receive information from

- Cost management plan
- Scope and requirements information
- Project schedule
- Quality management plan
- Risk register
- Lessons learned register

Cost estimating worksheets are developed and then refined as periodically as needed.

Description

You can use the element descriptions in Table 3.9A to assist you in developing a cost estimating worksheet and the element descriptions in Table 3.9B to assist you in developing a bottom-up cost estimating worksheet.

TABLE 3.9A   Elements of a Cost Estimating Worksheet

Document Element	Description
ID	Unique identifier, such as the WBS ID or activity ID
Parametric estimates	
Cost variable	Enter the cost estimating driver, such as hours, square feet, gallons, or some other quantifiable measure. Example: Square feet
Cost per unit	Record the cost per unit. Example: $9.50
Number of units	Enter the number of units. Example: 36
Cost estimate	Multiply the number of units times the cost per unit to calculate the estimate. Example: $9.50 × 36 = $342
Analogous estimates	
Previous activity	Enter a description of the previous activity. Example: Build a 160 square foot deck.
Previous cost	Document the cost of the previous activity. Example: $5,000
Current activity	Describe how the current activity is different. Example: Build a 200 square foot deck.
Multiplier	Divide the current activity by the previous activity to get a multiplier. Example: 200/160 = 1.25
Cost Estimate	Multiply the cost for the previous activity by the multiplier to calculate the cost estimate for the current activity. Example: $5,000 × 1.25 = $6,250
Multi-point estimate (Beta distribution)	
Optimistic cost	Determine an optimistic cost estimate. Optimistic estimates assume all costs were identified and there won't be any cost increases in material, labor, or other cost drivers. Example: $4,000
Most likely cost	Determine a most likely cost estimate. Most likely estimates assume that there will be some cost fluctuations but nothing out of the ordinary. Example: $5,000
Pessimistic cost	Determine a pessimistic cost estimate. Pessimistic estimates assume there are significant risks that will materialize and cause cost overruns. Example: $7,500
Weighting equation	Weight the three estimates and divide. The most common method of weighting is the beta distribution, where c = cost: $cE = (cO + c4M + cP)/6$ Example: (4000 + 4(5000) + 7500)/6
Expected cost	Enter the expected cost based on the beta distribution. Example: $5,250

You can use the element descriptions in Table 3.9B to assist you in developing a bottom-up cost estimating worksheet.

TABLE 3.9B   Elements of a Bottom-up Cost Estimating Worksheet

Document Element	Description
ID	Unique identifier, such as the WBS ID or activity ID.
Labor hours	Enter the estimated effort hours.
Labor rate	Enter the hourly of daily rate.
Total labor	Multiply the labor hours times the labor rate.
Material	Enter quotes for material, either from vendors or multiply the amount of material times the cost per unit.
Supplies	Enter quotes for supplies, either from vendors or multiply the amount of material times the cost per unit.
Equipment	Enter quotes to rent, lease, or purchase equipment.
Travel	Enter quotes for travel.
Other direct costs	Enter any other direct costs and document the type of cost.
Indirect costs	Enter any indirect costs, such as overhead.
Reserve	Document contingency reserve amounts, if any.
Estimate	The sum of the of labor, materials, supplies, equipment, travel, other direct costs, indirect costs, and any reserve costs

COST ESTIMATING WORKSHEET

Project Title: _____ **Date Prepared:** _____

Parametric Estimates

ID	Cost Variable	Cost per Unit	Number of Units	Cost Estimate

Analogous Estimates

ID	Previous Activity	Previous Cost	Current Activity	Multiplier	Cost Estimate

Multi-Point Estimates

ID	Optimistic Cost	Most Likely Cost	Pessimistic Cost	Weighting Equation	Expected Cost Estimate

BOTTOM-UP COST ESTIMATING WORKSHEET

Project Title: _____ **Date Prepared:** _____

ID	Labor Hours	Labor Rate	Total Labor	Material	Sup-plies	Equip-ment	Travel	Other Direct Costs	Indirect Costs	Reserve	Estimate

3.10 RESPONSIBILITY ASSIGNMENT MATRIX

The responsibility assignment matrix (RAM) shows the intersection of work packages and team members. RAMs can indicate different types of participation depending on the needs of the project. Some common types include

- Accountable
- Responsible
- Consulted
- Resource
- Informed
- Sign-off

The RAM always should include a key that explains what each of the levels of participation entails. An example follows using a RACI (responsible, accountable, consulted and informed) chart. The needs of your project should determine the fields for the RAM you use. Note that with a RACI chart, that each work package should have an accountable party, and only one accountable party. For a RACI chart, accountability is not shared.

The responsibility assignment matrix can receive information from

- Scope information
- Requirements documentation
- Resource management plan
- Stakeholder register

It is developed during planning and is progressively elaborated as more information about the scope and the resource requirements is known.

Tailoring Tips

Consider the following tips to help tailor the RAM to meet your needs:

- Tailor the types of participation appropriate for your project. Some projects require "sign-off" of specific deliverables, whereas others use the term "approve."
- Determine the appropriate level to record information on the RAM. Large projects with multiple vendors and large deliverables often use the RAM as the intersection of the WBS and the OBS (organizational breakdown structure). Small projects may use it at the deliverable or activity level to help enter schedule information.

Alignment

The RAM should be aligned and consistent with the following documents:

- Scope information
- Requirements documentation
- Procurement documents (RFP, RFQ, etc.)

Description

You can use the element descriptions in Table 3.10 to assist you in developing a responsibility assignment matrix.

TABLE 3.10   Elements of a Responsibility Assignment Matrix

Document Element	Description
Work package	Name of the work package you are assigning resources to. The RAM can be used at the work package level, control account level, or activity level.
Resource	Identify the person, division, or organization that will be working on the project.

RESPONSIBILITY ASSIGNMENT MATRIX

Project Title: _____ **Date Prepared:** _____

	Person 1	Person 2	Person 3	Person 4	Person N
Work Package 1	R	C	A		
Work Package 2		A		I	R
Work Package 3		R	R	A	
Work Package 4	A	R	I	C	
Work Package 5	C	R	R		A

R = Responsible: The person performing the work.

C = Consult: The person who has information necessary to complete the work.

A = Accountable: The person who is answerable to the project manager that the work is done on time, meets requirements, and is acceptable.

I = Inform: This person should be notified when the work is complete.

3.11 TEAM CHARTER

The team charter (also known as a team operating agreement) is used to establish ground rules and guidelines for the team. It is particularly useful on virtual teams and teams that are composed of members from different organizations. Using a team charter can help establish expectations and agreements on working effectively together. The contents of the team charter typically include

- Team values and principles
- Meeting guidelines
- Communication guidelines
- Decision-making process
- Conflict resolution process
- Team agreements

The team charter is generally developed once and changes if there is substantial team member turnover. The team should periodically revisit the team charter and reaffirm or update it accordingly.

Tailoring Tips

Consider the following tips to help tailor the team charter to meet your needs:

- If you bring in contractors for key roles in the project, you should include them in developing the team charter.
- If your organization has organizational values, make sure your team charter is aligned with the organizational values.
- International teams may need to spend more time developing this document, as different cultures have different ways of making decisions and resolving conflicts.
- Teams that use an Agile approach may call this document team norms or team working agreements. For Agile teams, this document focuses on how the team will self-organize and work together.

Alignment

The team charter should be aligned and consistent with the following documents:

- Resource management plan

Description

You can use the element descriptions in Table 3.11 to assist you in developing a team charter.

TABLE 3.11   Elements of a Team Charter

Document Element	Description
Team values and principles	List values and principles that the team agrees to operate within. Examples include mutual respect, operating from fact not opinion, etc.
Meeting guidelines	Identify guidelines that will keep meetings productive. Examples include decision makers must be present, start on time, stick to the agenda, etc.
Communication guidelines	List guidelines used for effective communication. Examples include everyone voices their opinion, not dominating the conversation, no interrupting, not using inflammatory language, etc.
Decision-making process	Describe the process used to make decisions. Indicate the relative power of the project manager for decision making as well as any voting procedures. Also indicate the circumstances under which a decision can be revisited.
Conflict resolution process	Describe the process for managing conflict, when a conflict will be escalated, when it should be tabled for later discussion, etc.
Team norms	Agreements about how the team will operate.
Other agreements	List any other agreements or approaches to ensuring a collaborative and productive working relationship among team members.

TEAM CHARTER

Project Title: _____ **Date Prepared:** _____

Team Values and Principles

1.

2.

3.

4.

5.

Meeting Guidelines

1.

2.

3.

4.

5.

Communication Guidelines

1.

2.

3.

4.

5.

Decision-Making Process

TEAM CHARTER

Conflict Resolution Process

Team Norms

Other Agreements

3.12 PROBABILITY AND IMPACT ASSESSMENT

There are three aspects of a probability and impact (PxI) assessment. The first entails the definitions for probability and impact. These are likely defined in the risk management plan, but if your project doesn't have a risk management plan, you can develop a PxI assessment to record definitions for the likelihood of events occurring (probability) and the impact on the various project objectives if they do occur.

The second aspect is a table that is used to plot each risk after performing a PxI assessment. The PxI assessment determines the probability and impact of the risk. The matrix provides a helpful way to view the various risks on the project and prioritize them for responses. It may be constructed for threats and opportunities. Information from this matrix will be transferred to the risk register.

Another benefit of the PxI matrix is that it provides an overview of the amount of risk on the project. The project team can get an idea of the overall project risk by seeing the number of risks in each square of the matrix. A project with many risks in the red zone will need more contingency to absorb the risk and likely more time and budget to develop and implement risk responses.

The third aspect of the assessment is determining which risks are considered low, medium, and high based on the combination of probability and impact.

The probability and impact assessment can receive information from

- Risk management plan
- Risk register

It provides information to the risk register.

The probability and impact assessment is developed once and does not usually change.

Tailoring Tips

Consider the following tips to help tailor the PxI matrix to meet your needs:

- On smaller projects, the impacts may be grouped together without distinguishing impact by objective.
- The matrix can be 3 × 3 for a small project, 5 × 5 for a medium project, and 10 × 10 for a complex or large project.
- To indicate the relative criticality of various objectives (usually scope, schedule, cost, and quality), you can include a tighter range of thresholds between levels. For example, if cost is a critical factor, consider a very low impact as a 2% variance, a low variance as a 4% impact, a medium variance as a 6% variance, a high variance as 8%, and a very high variance as a 10% variance. If the cost is more relaxed, you might have a loose range, such as very low impact as a 5% variance, a low variance as a 10% impact, a medium variance as a 15% variance, a high variance as 20%, and a very high variance as a 25% variance.
- The numbering structure of the probability and impact matrix can be tailored to emphasize the high risks by creating a nonlinear numbering structure. For example, impact scores can be set up to double every increment. For example: Very Low = .5, Low = 1, Medium = 2, High = 4, and Very High = 8.
- If there are other objectives that are important to the project, such as stakeholder satisfaction, you can incorporate them. Some organizations combine scope and quality into one objective called "technical" or "performance."
- You can make the assessment more robust by including urgency information to indicate how soon you need to implement the response for it to be effective.

- The combination of probability and impact that indicates a risk is high, medium, or low can be tailored to reflect the organization's risk appetite. An organization with a low risk appetite may rank events that fall in the medium or high range for both impact and probability as high risk. An organization with a higher risk threshold may only rank risk with a very high probability and impact as high risk.

Alignment

The probability and impact assessment should be aligned and consistent with the following documents:

- Risk management plan
- Risk register

Description

You can use the assessment descriptions in Table 3.12 to assist you in developing a probability and impact assessment.

TABLE 3.12   Elements of a Probability Impact Assessment

Document Element	Description	
Scope impact	Very High	The product does not meet the objectives and is effectively useless
	High	The product is deficient in multiple essential requirements
	Medium	The product is deficient in one major requirement or multiple minor requirements
	Low	The product is deficient in a few minor requirements
	Very Low	Minimal deviation from requirements
Quality impact	Very High	Performance is significantly below objectives and is effectively useless
	High	Major aspects of performance do not meet requirements
	Medium	At least one performance requirement is significantly deficient
	Low	There is minor deviation in performance
	Very Low	There is minimal deviation in performance
Schedule impact	Very High	Greater than 20% overall schedule increase
	High	Between 10 and 20% overall schedule increase
	Medium	Between 5 and 10% overall schedule increase
	Low	Noncritical paths have used all their float, or overall schedule increase of 1 to 5%
	Very Low	Slippage on noncritical paths but float remains
Cost impact	Very High	Cost increase of greater than 20%
	High	Cost increase of 10 to 20%
	Medium	Cost increase of 5 to 10%
	Low	Cost increase that requires use of all contingency funds
	Very Low	Cost increase that requires use of some contingency but some contingency funds remain

TABLE 3.12   Elements of a Probability Impact Assessment (*continued*)

Document Element	Description	
Probability	Very High	The event will most likely occur: 80% or greater probability
	High	The event will probably occur: 61 to 80% probability
	Medium	The event is likely to occur: 41 to 60% probability
	Low	The event may occur: 21 to 40% probability
	Very Low	The event is unlikely to occur: 1 to 20% probability
Risk rating	High	Any event with a probability of medium or above and a very high impact on any objective
		Any event with a probability of high or above and a high impact on any objective
		Any event with a probability of very high and a medium impact on any objective
		Any event that scores a medium on more than two objectives
	Medium	Any event with a probability of very low and a high or above impact on any objective
		Any event with a probability of low and a medium or above impact on any objective
		Any event with a probability of medium and a low to high impact on any objective
		Any event with a probability of high and a very low to medium impact on any objective
		Any event with a probability of very high and a low or very low impact on any objective
		Any event with a probability of very low and a medium impact on more than two objectives
	Low	Any event with a probability of medium and a very low impact on any objective
		Any event with a probability of low and a low or very low impact on any objective
		Any event with a probability of very low and a medium or less impact on any objective

PROBABILITY AND IMPACT RISK ASSESSMENT

Project Title: _____ Date Prepared: _____

Scope Impact

Very High	
High	
Medium	
Low	
Very Low	

Quality Impact

Very High	
High	
Medium	
Low	
Very Low	

Schedule Impact

Very High	
High	
Medium	
Low	
Very Low	

PROBABILITY AND IMPACT RISK ASSESSMENT

Cost Impact

Very High	
High	
Medium	
Low	
Very Low	

Probability

Very High	
High	
Medium	
Low	
Very Low	

Risk Rating

High	
Medium	
Low	

PROBABILITY AND IMPACT MATRIX

	Very Low	Low	Medium	High	Very High
Very High					
High					
Medium					
Low					
Very Low					

3.13 RISK DATA SHEET

A risk data sheet contains information about a specific identified risk. The information is filled in from the risk register and elaborated with more-detailed information. Typical information includes

- Risk identifier
- Risk description
- Status
- Risk cause
- Probability
- Impact on each objective
- Risk score
- Response strategies
- Revised probability
- Revised impact
- Revised score
- Responsible party
- Actions
- Secondary risks
- Residual risks
- Contingency plans
- Schedule or cost contingency
- Fallback plans
- Comments

The risk data sheet can receive information from the

- Risk register
- Probability and impact risk assessment

A risk data sheet is started as needed, and is continuously updated and elaborated throughout the life of the risk.

Tailoring Tips

Consider the following tips to help tailor the risk data sheet to meet your needs:

- Not all projects require risk data sheets. Where they are used, they are considered an extension of the risk register.
- For projects that use risk data sheets, not all risks require them. A risk data sheet is usually reserved for complex risks, high risks, or where there is a lot of information surrounding the risk that needs to be documented.
- You can add or delete any fields you feel necessary.

Alignment

The risk data sheet should be aligned and consistent with the following documents:

- Risk register
- Probability and impact assessment
- Risk report

Description

You can use the element descriptions in Table 3.13 to assist you in developing the risk data sheet.

TABLE 3.13   Elements of a Risk Data Sheet

Document Element	Description
Risk ID	Enter a unique risk identifier.
Risk description	Provide a detailed description of the risk.
Status	Enter the status as open or closed.
Risk cause	Describe the circumstances or drivers that are the source of the risk.
Probability	Determine the likelihood of the event or condition occurring.
Impact	Describe the impact on one or more of the project objectives.
Score	If you are using numeric scoring, multiply the probability times the impact to determine the risk score. If you are using relative scoring, then combine the two scores (e.g., high-low or medium-high).
Reponses	Describe the planned response strategy to the risk or condition.
Revised probability	Determine the likelihood of the event or condition occurring after the response has been implemented.
Revised impact	Describe the impact once the response has been implemented.
Revised score	Enter the revised risk score once the response has been implemented.
Responsible party	Identify the person responsible for managing the risk.
Actions	Describe any actions that need to be taken to respond to the risk.
Secondary risks	Describe new risks that arise out of the response strategies taken to address the risk.
Residual risk	Describe the remaining risk after response strategies are developed.
Contingency plan	Develop a plan that will be initiated if specific events occur, such as missing an intermediate milestone. Contingency plans are used when the risk or residual risk is accepted.
Contingency funds	Determine the funds needed to protect the budget from overrun.
Contingency time	Determine the time needed to protect the schedule from overrun.
Fallback plans	Devise a plan to use if other response strategies fail.
Comments	Provide any comments or additional helpful information about the risk event or condition.

RISK DATA SHEET

Project Title: _____ **Date Prepared:** _____

Risk ID:	Risk Description:						
Status:	Risk Cause:						

Probability	Impact				Score	Responses	
	Scope	Quality	Schedule	Cost			

Revised Probability	Revised Impact				Revised Score	Responsible Party	Actions
	Scope	Quality	Schedule	Cost			

Secondary Risks:

Residual Risk:

Contingency Plan:	Contingency Funds:
	Contingency Time:

Fallback Plans:

Comments:

3.14 PROCUREMENT STRATEGY

The procurement strategy is a project document that describes information about specific procurements. Typical information includes

- Delivery methods
- Contract types
- Procurement phases

The procurement strategy can receive information from

- Project charter
- Stakeholder register
- Project roadmap
- Requirements documentation
- Requirements traceability matrix
- Scope baseline
- Project schedule
- Resource management plan

It provides information to

- Project schedule
- Project budget
- Risk register

The procurement strategy is developed once for each procurement when needed.

Tailoring Tips

Consider the following tips to help tailor the procurement strategy to meet your needs:

- For a project that will be done using internal resources only, you do not need a procurement strategy.
- For projects with few procurements, consider combining this template with the procurement management plan.
- For simple purchases, or for purchases where you have worked with a vendor successfully for a length of time, you may not need a formal procurement strategy; rather, you would record the information in a statement of work that would be part of the contract.
- Work with the contracting or legal department to ensure compliance with organizational purchasing policies.

Alignment

The procurement strategy should be aligned and consistent with the following documents:

- Project charter
- Project roadmap
- Requirements documentation
- Requirements traceability matrix
- Schedule management plan

- Cost management plan
- Resource management plan
- Procurement management plan

Description

You can use the element descriptions in Table 3.14 to assist you in developing the procurement strategy.

TABLE 3.14   Elements of a Procurement Strategy

Document Element	Description	
Delivery methods	Professional services	Describe how the contractor will work with the buyer; for example, in a joint venture, as a representative, with or without subcontracting allowed.
	Construction services	Describe the limitations of delivery, such as design build, design bid build, etc.
Contract types	Describe the contract type, fixed, incentive, or award fees. Include the criteria associated with the fees. Common contract types include **Fixed Price** FFP – Firm Fixed Price FPIF – Fixed Price with Incentive Fee FP-EPA – Fixed Price with Economic Price Adjustment **Cost Reimbursable** CPFF – Cost Plus Fixed Fee CPIF – Cost Plus Incentive Fee CPAF – Cost Plus Award Fee **Time and Materials** (T&M)	
Procurement phases	List the procurement phases, milestones, criteria to advance to the next phase, and tests or evaluations for each phase. Include any knowledge transfer requirements.	

PROCUREMENT STRATEGY

Project Title: _____ **Date:** _____

Delivery Method

| |
| |
| |

Contract Type

☐ FFP	☐ FPIF	☐ FP-EPA	☐ CPFF	☐ CPIF	☐ CPAF	☐ T&M	☐ Other

Incentive or Award Fee	Criteria

Procurement Life Cycle

Phase	Entry Criteria	Key Deliverables or Milestones	Exit Criteria	Knowledge Transfer

3.15 SOURCE SELECTION CRITERIA

Source selection criteria is a set of attributes desired by the buyer that a seller must meet or exceed to be selected for a contract. The source selection criteria template is an aid in determining and rating the criteria that will be used to evaluate bid proposals. This is a multistep process.

1. The criteria to evaluate bid responses are determined.
2. A weight is assigned to each criterion. The sum of all the criteria must equal 100%.
3. The range of ratings for each criterion is determined, such as 1–5 or 1–10.
4. The performance necessary for each rating is defined.
5. Each proposal is evaluated against the criteria and is rated accordingly.
6. The weight is multiplied by the rate, and a score for each criterion is derived.
7. The scores are totaled, and the highest total score is the winner of the bid.

Evaluation criteria commonly include such items as

- Capability and capacity
- Product cost and life cycle cost
- Delivery dates
- Technical expertise
- Prior experience
- Proposed approach and work plan
- Key staff qualifications, availability, and competence
- Financial stability
- Management experience
- Training and knowledge transfer

The source selection criteria are established for each major procurement.

Tailoring Tips

Consider the following tips to help tailor the source selection criteria to meet your needs:

- For a small procurement that is not complex, you likely won't need to develop a weighted source selection criteria template.
- For international procurements you may also want to include familiarity with local laws and regulations, as well as experience and relationships in the locations involved.
- For construction projects, or projects with many logistical concerns, you could include logistics handling as a selection criterion.

Alignment

The source selection criteria should be aligned and consistent with the following documents:

- Requirements documentation
- Scope baseline
- Project schedule
- Resource management plan
- Procurement management plan
- Procurement strategy

Description

You can use the element descriptions in Table 3.15 to assist you in developing the source selection criteria.

TABLE 3.15   Elements of Source Selection Criteria

Document Element	Description
Criteria	1 Describe what a 1 means for the criteria. For example, for experience, it may mean that the bidder has no prior experience.
	2 Describe what a 2 means for the criteria. For example, for experience, it may mean that the bidder has done 1 similar job.
	3 Describe what a 3 means for the criteria. For example, for experience, it may mean that the bidder has done 3 to 5 similar jobs.
	4 Describe what a 4 means for the criteria. For example, for experience, it may mean that the bidder has done 5 to 10 similar jobs.
	5 Describe what a 5 means for the criteria. For example, for experience, it may mean that the job is the bidder's core competency.
Weight	Enter the weight for each criterion. Total weight for all criteria must equal 100%.
Candidate rating	Enter the rating per the criteria above.
Candidate score	Multiply the weight times the rating.
Total	Sum the scores for each candidate.

SOURCE SELECTION CRITERIA

Project Title: _____   **Date Prepared:** _____

	1	2	3	4	5
Criterion 1					
Criterion 2					
Criterion 3					

	Weight	Candidate 1 Rating	Candidate 1 Score	Candidate 2 Rating	Candidate 2 Score	Candidate 3 Rating	Candidate 3 Score
Criterion 1							
Criterion 2							
Criterion 3							
Totals							

3.16 STAKEHOLDER ANALYSIS

A stakeholder analysis is used to classify stakeholders. It can be used to help fill in the stakeholder register. Analyzing stakeholders can also help in planning stakeholder engagement for groups of stakeholders.

The following example is used to assess the relative power (high or low), the relative interest (high or low), and the attitude (friend or foe). There are many other ways to categorize stakeholders. Some examples include

- Influence/impact
- Power/urgency/legitimacy

Stakeholder analysis receives information from

- Project charter
- Procurement documents

A stakeholder analysis is started at the beginning of a project and updated as stakeholders come and go, and as more information about stakeholders is learned.

Tailoring Tips

Consider the following tips to help you tailor the stakeholder analysis to meet your needs:

- For projects with relatively homogenous stakeholders, you can use a 2 × 2 grid that only considers two variables, such as interest and influence.
- For larger projects consider using a 3 × 3 stakeholder cube. Tailor the categories to reflect the importance of various stakeholder variables.
- For smaller projects you can combine the stakeholder analysis with the stakeholder register.

Alignment

The stakeholder analysis should be aligned and consistent with the following documents:

- Stakeholder register
- Stakeholder engagement plan

Description

You can use the element descriptions in Table 3.16 to assist you in developing a stakeholder analysis.

TABLE 3.16   Stakeholder Analysis

Document Element	Description
Name or role	The stakeholder name, organization, or group
Interest	The level of concern the stakeholder has for the project
Influence	The degree to which the stakeholder can drive or influence outcomes for the project
Attitude	The degree to which the stakeholder supports the project

STAKEHOLDER ANALYSIS

Project Title: _____ **Date Prepared:** _____

Name or Role	Interest	Influence	Attitude

3.17 USER STORY

A user story is a brief description of a desired outcome, documented from a stakeholder's perspective. User stories keep stakeholder needs visible throughout the development process.

A user story contains the following information:

- Stakeholder
- Need/want
- Benefit

User stories generally follow this format:

As a *(stakeholder)*, I want *(want or need)* so I can *(benefit)*.

User stories may be kept on a backlog and prioritized for development by the product owner.

User stories can receive information from

- Vision statement
- Startup canvas
- Requirements documentation
- Stakeholder register

They provide information to

- Backlog

User stories are developed at the start of the project updated throughout the project as stakeholder needs and wants are clarified and evolve.

Tailoring Tips

Consider the following tips to help tailor the user stories to meet your needs:

- User stories can be documented on index cards, or in a software specifically designed for Agile projects.
- You can add space for story point estimates, the iteration the user story is assigned to, the responsible party, and completion criteria (AKA definition of done).

Alignment

User stories should be aligned and consistent with the following documents:

- Requirements documentation
- Scope information

Description

You can use the element descriptions in Table 3.17 to assist you in developing the retrospective.

TABLE 3.17   Elements of a Starfish Retrospective

Document Element	Description
Stakeholder	The person or group who has a need or requirement. The stakeholder can be a role, position, or the name of a specific stakeholder.
Need	A description of what they want or they need a feature or function to do.
Benefit	A description of the value that the feature or function will provide.

USER STORY

As a _____

I want _____

So I can _____

3.18 RETROSPECTIVE

A retrospective is an activity that is performed at the end of every iteration (sprint). It is a safe environment to explore things that worked in the previous iteration, and things that can be improved. Retrospectives keep the team aligned with continuous improvement.

Information is usually recorded on sticky notes or recorded in software for remote teams. There are several common retrospective approaches. The starfish retrospective collects the following information:

- Start
- Stop
- Keep
- More
- Less

Another common retrospective is the 4 Ls where team members document the following information:

- Liked
- Learned
- Lacked
- Longed for

There are many retrospective approaches, but regardless of the approach, the intent of a retrospective is to engage the team in improving their performance so they become more efficient in each subsequent iteration.

Alignment

The retrospective should be aligned and consistent with the following documents:

- Lessons learned summary
- Project closeout

Description

You can use the element descriptions in Table 3.18 to assist you in developing the retrospective.

TABLE 3.18   Elements of a Starfish Retrospective

Document Element	Description
Start	Actions and behaviors that the team will begin to implement
Stop	Actions or behaviors that the team will cease doing
Keep	Practices that the team should continue with
More	Practices that were not done consistently that should be done more often
Less	Practices that were done too much or that should be reduced

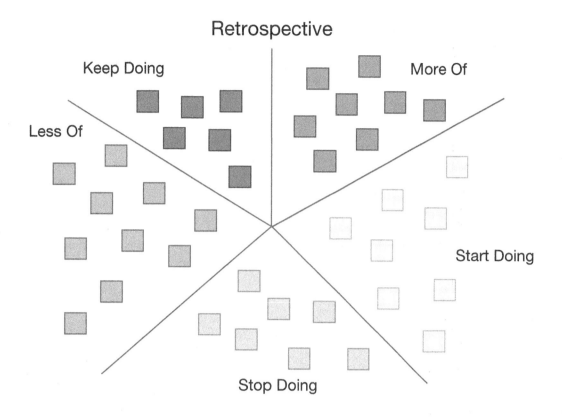

RETROSPECTIVE

Project Title: _____   **Date Prepared:** _____

Liked	Lacked	Learned	Longed For

Logs and Registers

Logs and registers are very helpful in keeping track of changing information. They are dynamic documents that are started at the beginning of the project and are kept up to date throughout the project.

Most of the templates in this chapter are useful for waterfall as well as adaptive projects. Examples include an assumption log, decision log, and issue log – though the issue log may be known as an impediment log in projects that use an Agile methodology. A backlog is used primarily for Agile projects, and a change log is used primarily for predictive projects. There are eight log/register templates:

- Assumption log
- Backlog
- Change log
- Decision log
- Issue log
- Stakeholder register
- Risk register
- Lessons learned register

Several of the templates in this section can be combined into one document, especially if you create the templates in a spreadsheet, with each worksheet containing information for one log. For example, a RAID log contains:

- Risk log
- Assumption log
- Issue log
- Decision log

You may find that several of the logs interact, for example, an assumption that is not validated may become a risk. A risk that occurs may be considered an issue. In these cases, it is best to close the item in one log with a notation that it has moved to a different log. In the notes section of the new log it is useful to note if the item came from a different log or register.

4.1 ASSUMPTION LOG

Assumptions are factors that are considered to be true but without proof. Constraints are also documented in this log. Constraints are limiting factors that affect the execution of the project. Typical constraints include a predetermined budget or fixed milestones for deliverables. Information in the assumption log includes

- Identifier
- Category
- Assumption or constraint
- Responsible party
- Due date
- Actions
- Status
- Comments

Assumptions can come from any document in the project. They can also be determined by the project team. Constraints may be documented in the project charter and are determined by the customer, sponsor, or regulatory agencies.

The assumption log provides information to

- Requirements documentation
- Project scope statement
- Duration estimates
- Project schedule
- Quality management plan
- Resource estimates
- Risk register
- Stakeholder engagement plan

The assumption log is a dynamic document that is updated throughout the project. Assumptions are progressively elaborated throughout the project, and when they are eventually validated, they are no longer assumptions.

Tailoring Tips

Consider the following tips to help you tailor the assumption log to meet your needs:

- Combine the assumption log with the issue register and the decision log to create an AID Log (A = assumption, I = issue, D = decision). You can create them in a spreadsheet with each sheet dedicated to either assumptions, issues, or decisions.
- If you have a very large project, you may want to keep the constraints in a separate log from the assumptions.

Alignment

The assumption log should be aligned and consistent with the following documents:

- Project charter
- Project brief
- Issue log
- Risk register

Description

You can use the element descriptions in Table 4.1 to assist you in developing the assumption log.

TABLE 4.1   Elements of an Assumption Log

Document Element	Description
ID	Identifier
Category	The category of the assumption or constraint
Assumption/constraint	A description of the assumption or constraint
Responsible party	The person who is tasked with following up on the assumption to validate if it is true or not
Due date	The date by which the assumption needs to be validated
Actions	Actions that need to be taken to validate assumptions
Status	The status of the assumptions, such as active, transferred, or closed
Comments	Any additional information regarding the assumption or constraint

ASSUMPTION LOG

Project Title: _____   **Date Prepared:** _____

ID	Category	Assumption/ Constraint	Respon- sible Party	Due Date	Actions	Status	Comments

4.2 BACKLOG

A backlog is used when a project uses an adaptive approach, such as Agile. It is used to prioritize work. A backlog is developed at the very beginning of a project, often in conjunction with the product vision statement. The backlog is used to enter all requirements so they can be prioritized.

The product backlog includes at least

- ID
- Summary description
- Priority
- Status

The backlog is developed at the start of the project and is updated throughout the project.

Tailoring Tips

Consider the following tips to help tailor the product backlog to meet your needs:

- Rather than using requirements, you can have user stories in the backlog. Alternatively, you can use the backlog for requirements and have a column that indicates the user story each requirement is associated with.
- To provide more detail, you can indicate which iteration or release a requirement will be incorporated into.
- You may want to indicate the user type that will benefit from the requirement, such as customer, administrator, manager, etc.
- For large projects, it helps to categorize requirements, so having a column that indicates the category can be useful.

Alignment

The backlog should be aligned and consistent with the following documents:

- Product vision
- Roadmap
- Release plan

Description

You can use the element descriptions in Table 4.2 to assist you in developing the product backlog.

TABLE 4.2   Elements of a Product Backlog

Document Element	Description
ID	A unique identifier
Summary description	A brief description of the requirement or need. The description should be no more than one or two sentences.
Priority	A way of prioritizing or ranking the requirements. This can be in summary groups, such as high, medium, and low, or it can be numbered 1, 2, 3.
Status	Indicates if the requirement is not started, in progress, or complete.

BACKLOG

Project Title: _____ **Date Prepared:** _____

ID	Requirement	Priority	Status

4.3 CHANGE LOG

The change log is used to track changes from the change request through the final decision. Typical information includes

- Identifier
- Category
- Description
- Requestor
- Submission date
- Status
- Disposition

The change log is related to the

- Change request
- Change management plan

The change log is a dynamic document that is updated throughout the project.

Tailoring Tips

Consider the following tips to help tailor the change log to meet your needs:

- You can include additional summary information from the change request in the log, such as cost or schedule impact.
- You can add a check box that indicates whether the change is mandatory (such as a legal requirement) or discretionary.
- The change log can also record information to track configuration management, such as which configurable items are impacted.
- Some IT projects include a field that indicates if a change is a bug fix.

Alignment

The change log should be aligned and consistent with the following documents:

- Change management plan
- Change request

Description

You can use the element descriptions in Table 4.3 to assist you in developing the change log.

TABLE 4.3   Elements of a Change Log

Document Element	Description
Identifier	Enter a unique change identifier.
Category	Enter the category from the change request form.
Description	Describe the proposed change.
Requestor	Enter the name of the person requesting the change.
Submission date	Enter the date the change was submitted.
Status	Enter the status as open, pending, closed.
Disposition	Enter the outcome of the change request as approved, deferred, or rejected.

CHANGE LOG

Project Title: _____   **Date Prepared:** _____

ID	Category	Description of Change	Requestor	Date	Status	Disposition

4.4 DECISION LOG

Frequently, there are alternatives in developing a product or managing a project. Using a decision log can help keep track of the decisions that were made, who made them, and when they were made. A decision log can include

- Identifier
- Category
- Decision
- Responsible party
- Date
- Comments

The decision log can be very helpful in managing the day-to-day activities of the project. It is a dynamic document that is created at the start of the project and is maintained throughout the project.

Tailoring Tips

Consider the following tips to help tailor the decision log to meet your needs:

- For projects that are large, complicated, or complex, you can add fields to identify the impacts of the decision on deliverables or project objectives.
- You could add a field that documents which stakeholders are impacted by the decision, should be involved with making the decision, or should be informed of the decision.

Alignment

The decision log should be aligned and consistent with the following documents:

- Project scope statement
- Responsibility assignment matrix
- Communications management plan
- Issue register

Description

You can use the element descriptions in Table 4.4 to assist you in developing the decision log.

TABLE 4.4   Elements of a Decision Log

Document Element	Description
ID	Enter a unique decision identifier.
Category	Document the type of decision, such as technical, project, process, etc.
Decision	Provide a detailed description of the decision.
Responsible party	Identify the person authorized to make the decision.
Priority	Enter the date the decision was made and authorized.
Comments	Enter any further information to clarify the decision, alternatives considered, the reason the decision was made, and the impact of the decision.

DECISION LOG

Project Title: _____   **Date Prepared:** _____

ID	Category	Decision	Responsible Party	Date	Comments

4.5 ISSUE LOG

The issue log is used to record and track issues. An issue is defined as a current condition or situation that could have an impact on the project objectives. Examples of issues are points or matters in question that are in dispute or under discussion, or over which there are opposing views or disagreements. Issues can also arise from a risk event that has occurred and must now be dealt with. An issue log includes

- Identifier
- Type
- Issue description
- Priority
- Impact on objectives
- Responsible party
- Status
- Resolution date
- Final resolution
- Comments

The issue log is a dynamic document that is created at the start of the project and is maintained throughout the project.

Tailoring Tips

Consider the following tips to help tailor the issue log to meet your needs:

- You may want to add information on the source of the issue.
- You could add a field that documents which stakeholders are impacted by the issue or should be involved with resolving the issue.
- Hybrid projects, or projects that use Agile methods, may use the term blocker or impediment when referring to issues.

Alignment

The issue log should be aligned and consistent with the following documents:

- Risk register
- Decision log
- Lessons learned register

Description

You can use the element descriptions in Table 4.5 to assist you in developing the issue log.

TABLE 4.5   Elements of an Issue Log

Document Element	Description
ID	Enter a unique issue identifier.
Type	Document the type or category of the issue, such as stakeholder issue, technical issue, conflict, etc.
Issue description	Provide a detailed description of the issue.
Priority	Define the priority, such as urgent, high, medium, or low.
Impact on objectives	Identify the project objectives that the issue impacts and the degree of impact.
Responsible party	Identify the person who is assigned to resolve the issue.
Status	Denote the status of the issue as open or closed.
Resolution date	Document the date by which the issue needs to be resolved.
Final resolution	Describe how the issue was resolved.
Comments	Document any clarifying comments about the issue, resolution, or other fields on the form.

ISSUE LOG

Project Title: _____ **Date Prepared:** _____

ID	Type	Issue Description	Priority	Impact on Objectives

Responsible Party	Status	Res. Date	Final Resolution	Comments

4.6 STAKEHOLDER REGISTER

The stakeholder register is used to identify those people and organizations impacted by the project and to document relevant information about each stakeholder. Relevant information can include

- Name
- Position in the organization
- Role in the project
- Contact information
- List of stakeholder's major requirements
- List of stakeholder's expectations
- Classification of each stakeholder

Initially, you will not have enough information to complete the stakeholder register. As the project gets under way, you will gain additional information and understanding about each stakeholder's requirements, expectations, and classification and the stakeholder register will become more robust.

The stakeholder register receives information from

- Project charter
- Project startup canvas
- Procurement documents

It is related to stakeholder analysis.

It provides information to

- Requirements documentation
- Quality management plan
- Communications management plan
- Risk management plan
- Risk register
- Stakeholder engagement plan

The stakeholder register is a dynamic project document. The stakeholders, their level of influence, requirements, and classification are likely to change throughout the project.

Tailoring Tips

Consider the following tips to help you tailor the stakeholder register to meet your needs:

- Combine the position in the organization with the role on the project, especially if it is a smaller project and everyone knows everyone else's position.
- Combine the stakeholder analysis information with the stakeholder register.
- Combine with the communications management plan for small projects.
- Eliminate position, role, and contact information for small internal projects.

Alignment

The stakeholder register should be aligned and consistent with the following documents:

- Project charter
- Stakeholder analysis
- Stakeholder engagement plan

Description

You can use the element descriptions in Table 4.6 to assist you in developing the stakeholder register.

TABLE 4.6   Elements of a Stakeholder Register

Document Element	Description
Name	Stakeholder's name. If you don't have a name you can substitute a position or organization until you have more information
Position/Role	The position and/or role the stakeholder holds in the organization. Examples of positions include programmer, human resources analyst, or quality assurance specialist. Roles indicate the function the stakeholder performs on the project team, such as testing lead, Scrum Master, or scheduler.
Contact information	How to communicate with the stakeholder, such as their phone number, email address, or physical address
Requirements	High-level needs for the project and/or product
Expectations	Main expectations of the project and/or product
Classification	Some projects may categorize stakeholders as friend, foe, or neutral; others may classify them as high, medium, or low impact.

STAKEHOLDER REGISTER

Project Title: _____

Date Prepared: _____

Name	Position/Role	Contact Information	Requirements	Expectations	Classification

---

4.7 RISK REGISTER

The risk register captures the details of individual risks. It documents the results of risk analysis, risk response planning, response implementation, and current status. It is used to track information about identified risks over the course of the project. Typical information includes

- Risk identifier
- Risk statement
- Risk owner
- Probability of occurring
- Impact on objectives if the risk occurs
- Risk score
- Response strategies
- Revised probability
- Revised impact
- Revised score
- Actions
- Status
- Comments

The risk register can receive information from anywhere in the project environment, including startup documents. Some documents that should be specifically reviewed for input include

- Assumption log
- Issue log
- Lessons learned register
- Requirements management plan
- Requirements documentation
- Scope information
- Schedule management plan
- Duration estimates
- Schedule
- Cost management plan
- Cost estimates
- Budget
- Quality management plan
- Resource management plan
- Risk management plan
- Procurement documents
- Contracts
- Stakeholder register

The risk register provides information to

- Scope statement
- Duration estimates
- Cost estimates
- Quality management plan
- Resource requirements
- Risk report
- Procurement management plan

- Stakeholder engagement plan
- Lessons learned register
- Project closeout

The risk register is developed at the start of the project and is updated throughout the project.

Tailoring Tips

Consider the following tips to help tailor the risk register to meet your needs:

- You may want to add information on the source of the risk.
- Hybrid projects or projects that use Agile methods may use the term blocker or impediment when referring to risks.

Alignment

The risk register should be aligned and consistent with the following documents:

- Project budget
- Issue log
- Assumption log
- Decision log
- Lessons learned register

Description

You can use the element descriptions in Table 4.7 to assist you in developing the risk register.

TABLE 4.7   Elements of a Risk Register

Document Element	Description
Risk ID	Enter a unique risk identifier.
Risk statement	Describe the risk event or condition. A risk statement is usually phrased as "EVENT may occur, causing IMPACT" or "If CONDITION exists, EVENT may occur, leading to EFFECT."
Risk owner	The person responsible for managing and tracking the risk
Probability	Determine the likelihood of the event or condition occurring.
Impact	Describe the impact on one or more of the project objectives.
Score	If you are using numeric scoring, multiply the probability times the impact to determine the risk score. If you are using relative scoring, then combine the two scores (e.g., high-low or medium-high).
Response	Describe the planned response strategy to the risk or condition.
Revised probability	Determine the likelihood of the event or condition occurring after the response has been implemented.
Revised impact	Describe the impact once the response has been implemented.
Revised score	Enter the revised risk score once the response has been implemented.
Actions	Describe any actions that need to be taken to respond to the risk.
Status	Enter the status as open or closed.
Comments	Provide any comments or additional helpful information about the risk event or condition.

RISK REGISTER

Project Title: _____   **Date Prepared:** _____

ID	Risk Statement	Owner	Proba-bility	Impact				Score	Response
				Scope	Quality	Schedule	Cost		

Revised Probability	Impact				Revised Score	Respon-sible Party	Actions	Status	Com-ments
	Scope	Quality	Schedule	Cost					

4.8 LESSONS LEARNED REGISTER

The lessons learned register is used to record challenges, problems, good practices, and other information that can be used in the current project, passed along to other projects to avoid repeating mistakes. It can be shared with the organization to improve organizational processes and procedures. Lessons learned can be project oriented or product oriented. They can include information on risks, issues, procurements, quality defects, and any areas of poor or outstanding performance. A lessons learned register includes

- Identifier
- Category
- Trigger
- Lesson
- Responsible party
- Comments

The lessons learned register is a dynamic document that is updated throughout the project.

Tailoring Tips

Consider the following tips to help tailor the lessons learned register to meet your needs:

- You can add information on the person identifying the lesson, especially if the person identifying the lesson and the person accountable for implementing it are different.
- Information on the next implementation opportunity and the expected implementation date can be used to ensure that the information isn't just recorded, but it is acted on as well.
- You can add a check box to indicate whether the lesson impacts an organizational system, policy, or practice, or whether it can be implemented without the need to escalate up through the organization.

Alignment

The lessons learned register should be aligned and consistent with the following documents:

- Change management plan
- Change log
- Issue log
- Decision log
- Lessons learned summary

Description

You can use the element descriptions in Table 4.8 to assist you in developing the lessons learned register.

TABLE 4.8   Elements of a Lessons Learned Register

Document Element	Description
ID	Enter a unique lesson identifier.
Category	Document the category of lesson, such as process, technical, environmental, stakeholder, phase, etc.
Trigger	Describe the context, event, or condition that led to the challenge, problem, or beneficial outcome.
Lesson	Articulate the lesson that can be passed on to other projects and to the organization.
Responsible party	Identify the person who is assigned to implement any changes to ensure the lesson is communicated and distributed.
Comments	Document any clarifying comments about the challenge, problem, good practice, or other fields on the form.

LESSONS LEARNED REGISTER

Project Title: _____

Date Prepared: _____

ID	Category	Trigger	Lesson	Responsible Party	Comments

5

Reports and Audits

All projects need to report project status; however, projects that use predictive approaches are more likely to use reports to communicate the status of their project. For smaller projects, team progress reports and project status reports are sufficient. Larger projects might use an earned value report and a risk report.

Not all projects are subject to audits. Those that are, tend to be larger projects. Audits may be conducted at the project level, or they may be done by the project management office (PMO) or at a program or portfolio level.

There are 12 templates associated with reports and audits:

- Team progress report
- Project status report
- Variance analysis report
- Earned value report
- Risk report
- Contractor status report
- Contract closeout report
- Lessons learned report
- Project closeout report
- Quality audit
- Risk audit
- Procurement audit

Progress and status reports follow a set cadence, such as monthly. Closeout and lessons learned reports are used as needed. It is unlikely that you will need to change format of the reports and audits during the project.

5.1 TEAM MEMBER PROGRESS REPORT

The team member progress report is filled out by team members and submitted to the project manager on a regular basis. It tracks schedule, quality, and cost status for the current reporting period and provides planned information for the next reporting period. Progress reports also identify new risks and issues that have arisen in the current reporting period. Typical information includes

- Activities planned for the current reporting period
- Activities accomplished in the current reporting period
- Activities planned but not accomplished in the current reporting period
- Root causes of activities variances
- Funds spent in the current reporting period
- Funds planned to be spent for the current reporting period
- Root causes of funds variances
- Quality variances identified in the current reporting period
- Planned corrective or preventive action
- Activities planned for the next reporting period
- Costs planned for the next reporting period
- New risks identified
- New issues identified
- Comments

This information is generally compiled by the project manager into a project status report. The team member progress report is submitted at predefined intervals throughout the project.

Tailoring Tips

Consider the following tips to tailor the team member progress report to meet your needs:

- You can add a field for escalations to identify those areas that need to be escalated to the sponsor, program manager, or other appropriate individual.
- Some reports include a field to record decisions made. These would be transferred to the project decision log.
- If your organization has a robust knowledge management process, you might consider adding fields for knowledge transfer or lessons learned. These can then be transferred to the organization's knowledge repository or lessons learned register.

Alignment

The team member progress report should be aligned and consistent with the following documents:

- Project schedule
- Cost estimates
- Project budget
- Issue log
- Risk register
- Project status report
- Variance analysis
- Earned value status report

Description

You can use the element descriptions in Table 5.1 to assist you in developing the team member status report.

TABLE 5.1   Elements of a Team Member Status Report

Document Element	Description
Activities planned this reporting period	List all activities scheduled for this period, including work to be started, continued, or completed.
Activities accomplished this reporting period	List all activities accomplished this period, including work that was started, continued, or completed.
Activities planned but not accomplished this reporting period	List all activities that were scheduled for this period, but not started, continued, or completed.
Root cause of variances	For any work that was not accomplished as scheduled, identify the cause of the variance.
Funds spent this reporting period	Record funds spent this period.
Funds planned to be spent this reporting period	Record funds that were planned to be spent this period.
Root cause of variances	For any expenditures that were over or under plan, identify the cause of the variance. Include information on labor versus material variances. Identify if the basis of estimates or the assumptions were inaccurate.
Quality variances identified this period	Identify any product performance or quality variance.
Planned corrective of preventive action	Identify any actions needed to recover cost, schedule, or quality variances or prevent future variances.
Activities planned for next reporting period	List all activities scheduled for next period, including work to be started, continued, or completed.
Costs planned for next reporting period	Identify funds planned to be expended next period.
New risks identified	Identify any new risks that have arisen. New risks should be recorded in the risk register as well.
Issues	Identify any new issues that have arisen. New issues should be recorded in the issue log as well.
Comments	Document any comments that add relevance to this report.

TEAM MEMBER PROGRESS REPORT

Project Title: _____ Date Prepared: _____

Team Member _____ Role _____

Activities Planned for This Reporting Period

1.	
2.	
3.	
4.	

Activities Accomplished This Reporting Period

1.	
2.	
3.	
4.	

Activities Planned But Not Accomplished This Reporting Period

1.	
2.	
3.	
4.	

TEAM MEMBER STATUS REPORT

Root Cause of Activity Variances

Funds Spent This Period	Funds Planned to Be Spent This Period	Root Cause of Variances

Quality Variances Identified This Period	Planned Corrective or Preventive Action

TEAM MEMBER STATUS REPORT

Activities Planned for Next Reporting Period

1.	
2.	
3.	
4.	

Costs Planned for Next Reporting Period

New Risks Identified

New Issues Identified

Comments

5.2 PROJECT STATUS REPORT

The project status report (sometimes known as a performance report or progress report) is filled out by the project manager and submitted on a regular basis to the sponsor, project portfolio management group, PMO, or other project oversight person or group. The information is compiled from the team member progress reports and includes overall project performance. It contains summary-level information, such as accomplishments, rather than detailed activity-level information. The project status report tracks schedule and cost status for the current reporting period and provides planned information for the next reporting period. It indicates impacts to milestones and cost reserves as well as identifying new risks and issues that have arisen in the current reporting period. Typical information includes

- Accomplishments for the current period
- Accomplishments planned but not completed in the current period
- Root causes of accomplishment variances
- Impact to upcoming milestones or project due date
- Planned corrective or preventive action
- Funds spent in the current reporting period
- Root causes of budget variances
- Impact to overall budget or contingency funds
- Planned corrective or preventive action
- Accomplishments planned for the next reporting period
- Costs planned for the next reporting period
- New risks
- New issues
- Comments

The project status report is submitted at predefined intervals throughout the project.

Tailoring Tips

Consider the following tips to help tailor the project status report to meet your needs:

- You can add a field for escalations to identify those areas that need to be escalated to the sponsor, program manager, or other appropriate individuals.
- Some reports include a field to record decisions made. These would be transferred to the project decision log.
- If there were any change requests that were submitted during the reporting period, you may want to summarize them and refer the reader to the change log.
- If your organization has a robust knowledge management process, you might consider adding fields for knowledge transfer or lessons learned. These can then be transferred to the organization's knowledge repository or lessons learned register.
- In addition to tailoring the content of the project status report, you can tailor the presentation. Many PMOs have reporting software that transforms the data into dashboards, heat reports, stop light charts, or other representations.

Alignment

The project status report should be aligned and consistent with the following documents:

- Team member progress reports
- Project schedule

- Cost estimates
- Project budget
- Issue log
- Risk register
- Variance analysis
- Earned value status report
- Contractor status report

Description

You can use the element descriptions in Table 5.2 to assist you in developing the project status report.

TABLE 5.2   Elements of a Project Status Report

Document Element	Description
Accomplishments for this period	List all work packages or other accomplishments scheduled for completion for the current reporting period.
Accomplishments planned but not completed this period	List all work packages or other accomplishments scheduled for the current period but not completed.
Root cause of variances	Identify the cause of the variance for any work that was not accomplished as scheduled for the current period.
Impact to upcoming milestones or project due date	Identify any impact to any upcoming milestones or overall project schedule for any work that was not accomplished as scheduled. Identify any work currently behind on the critical path or if the critical path has changed based on the variance.
Planned corrective or preventive action	Identify any actions needed to make up schedule variances or prevent future schedule variances.
Funds spent this reporting period	Record funds spent this period.
Root cause of variance	Identify the cause of the variance for any expenditure over or under plan. Include information on the labor variance versus material variance and whether the variance is due to the basis of estimates or estimating assumptions.
Impact to overall budget or contingency funds	Indicate the impact to the overall project budget or whether contingency funds must be expended.
Planned corrective or preventive action	Identify any actions needed to recover cost variances or to prevent future schedule variances.
Accomplishments planned for next reporting period	List all work packages or accomplishments scheduled for completion next period.
Costs planned for next reporting period	Identify funds planned to be expended next period.
New risks	Identify any new risks that have been identified this period. These risks should be recorded in the risk register as well.
New Issues	Identify any new issues that have arisen this period. These issues should be recorded in the issue log as well.
Comments	Record any comments that add relevance to the report.

PROJECT STATUS REPORT

Project Title: _____ **Date Prepared:** _____

Project Manager _____ **Sponsor** _____

Accomplishments for This Period

1.	
2.	
3.	
4.	

Accomplishments Planned But Not Completed This Period

1.	
2.	
3.	
4.	

Root Cause of Variances

PROJECT STATUS REPORT

Impact to Upcoming Milestones or Project Due Date

Planned Corrective or Preventive Action

Funds Spent This Reporting Period

Root Cause of Variances

PROJECT STATUS REPORT

Impact to Overall Budget or Contingency Funds

Planned Corrective or Preventive Action

Accomplishments Planned for Next Reporting Period

1.

2.

3.

4.

PROJECT STATUS REPORT

Costs Planned for Next Reporting Period

New Risks

New Issues

Comments

5.3 VARIANCE ANALYSIS REPORT

Variance analysis reports collect and assemble information on project performance variances. Common topics are schedule, cost, and technical variances. Technical variances incorporate both scope and quality performance variances. Information included in a variance analysis includes

- Schedule variance
  - Planned results
  - Actual results
  - Variance
  - Root cause
  - Planned response
- Cost variance
  - Planned results
  - Actual results
  - Variance
  - Root cause
  - Planned response
- Technical variance
  - Planned results
  - Actual results
  - Variance
  - Root cause
  - Planned response

A variance analysis can be provided as a standalone report, as part of the project status report, or as backup to an earned value status report.

Tailoring Tips

Consider the following tips to help tailor the variance analysis to meet your needs:

- Technical variance can be decomposed into scope and quality variance.
- The variance analysis can be done at an activity, resource, work package, control account, or project level depending on your needs.
- You can add a check box to indicate if the information needs to be escalated to the sponsor, program manager, or other appropriate individuals.
- You may want to add a field that indicates the implications of continued variance. This can include a forecast based on a trend analysis or based on identified responses.
- In addition to tailoring the content of the variance analysis, you can tailor the presentation. Many PMOs have reporting software that transforms the data into dashboards, heat reports, stop light charts, or other representations.

Alignment

The variance analysis should be aligned and consistent with the following documents:

- Team member progress reports
- Project status report

- Project schedule
- Cost estimates
- Project budget
- Issue log
- Earned value status report
- Contractor status report

Description

You can use the element descriptions in Table 5.3 to assist you in developing the variance analysis.

TABLE 5.3   Elements of Variance Analysis

Document Elements	Description	
Schedule variance	Planned result	Describe the work planned to be accomplished.
	Actual result	Describe the work actually accomplished.
	Variance	Describe the variance.
	Root cause	Identify the root cause of the variance.
	Planned response	Document the planned corrective or preventive action.
Cost variance	Planned result	Record the planned costs for the work planned to be accomplished.
	Actual result	Record the actual costs expended.
	Variance	Calculate the variance.
	Root cause	Identify the root cause of the variance.
	Planned response	Document the planned corrective or preventive action.
Technical variance	Planned result	Describe the planned technical performance or metrics.
	Actual result	Describe the actual technical performance or metrics.
	Variance	Describe the variance.
	Root cause	Identify the root cause of the variance.
	Planned response	Document the planned corrective action.

VARIANCE ANALYSIS REPORT

Project Title: _____ **Date Prepared:** _____

Schedule Variance

Planned Result	Actual Result	Variance

Root Cause

Planned Response

Cost Variance

Planned Result	Actual Result	Variance

Root Cause

VARIANCE ANALYSIS REPORT

Planned Response

Technical Variance

Planned Result	Actual Result	Variance

Root Cause

Planned Response

5.4 EARNED VALUE ANALYSIS

Earned value analysis shows specific mathematical metrics that are designed to reflect the health of the project by integrating technical, schedule, and cost information. Information can be reported for the current reporting period and on a cumulative basis. Earned value analysis can also be used to forecast the total cost of the project at completion or the efficiency required to complete the project for the baseline budget. Information that is generally collected includes

- Budget at completion (BAC)
- Planned value (PV)
- Earned value (EV)
- Actual cost (AC)
- Schedule variance (SV)
- Cost variance (CV)
- Schedule performance index (SPI)
- Cost performance index (CPI)
- Percent planned
- Percent earned
- Percent spent
- Estimates at completion (EAC)
- To complete performance index (TCPI)

Earned value analysis information can be provided as a standalone report or as part of the project status report. Earned value analysis is conducted at pre-defined intervals throughout the project.

Tailoring Tips

Consider the following tips to help tailor the earned value analysis to meet your needs:

- The earned value analysis can be done at the control account and/or project level, depending on your needs.
- You may want to add a field that indicates the implications of continued variance. This can include a forecast based on a trend analysis or based on identified responses.
- Several different equations can be used to calculate the EAC, depending on whether the remaining work will be completed at the budgeted rate or at the current rate. Two options are presented on this template.
- There are options to calculate a TCPI. Use the information from your project to determine the best approach for reporting.
- You may want to add information that indicates the implications of continued schedule variance. This can include a schedule forecast using SPI as the basis for a trend analysis or based on analyzing the critical path.
- Some organizations are starting to embrace earned schedule metrics. You can update this form to include various earned schedule calculations in addition to a critical path analysis.
- In addition to tailoring the content of the earned value analysis, you can tailor the presentation. Many PMOs have reporting software that transforms the data into dashboards, control charts, S-curves, or other representations.

Alignment

Earned value analysis should be aligned and consistent with the following documents:

- Project status report
- Project schedule
- Project budget
- Variance analysis report
- Contractor status report

Description

You can use the element descriptions in Table 5.4 to assist you in developing an earned value analysis.

TABLE 5.4   Elements of Earned Value Analysis

Document Element	Description
Planned value	Enter the value of the work planned to be accomplished.
Earned value	Enter the value of the work actually accomplished.
Actual cost	Enter the cost for the work accomplished.
Schedule variance	Calculate the schedule variance by subtracting the planned value from the earned value. $SV = EV - PV$
Cost variance	Calculate the cost variance by subtracting the actual cost from the earned value. $CV = EV - AC$
Schedule performance index	Calculate the schedule performance index by dividing earned value by the planned value. $SPI = EV/PV$
Cost performance index	Calculate the cost performance index by dividing the earned value by the actual cost. $CPI = EV/AC$
Root cause of schedule variance	Identify the root cause of the schedule variance.
Schedule impact	Describe the impact on deliverables, milestones, or critical path.
Root cause of cost variance	Identify the root cause of the cost variance.
Budget impact	Describe the impact on the project budget, contingency funds and reserves, and any intended actions to address the variance.
Percent planned	Indicate the cumulative percent of the work planned to be accomplished. PV/BAC
Percent earned	Indicate the cumulative percent of work that has been accomplished. EV/BAC
Percent spent	Indicate the total costs spent to accomplish the work. AC/BAC
Estimates at completion	Determine an appropriate method to forecast the total expenditures at the project completion. Calculate the forecast and justify the reason for selecting the particular estimate at completion. For example:
	If the CPI is expected to remain the same for the remainder of the project: $EAC = BAC/CPI$
	If both the CPI and SPI will influence the remaining work: $EAC = AC + [(BAC - EV)/(CPI \times SPI)]$
To complete performance index	Calculate the work remaining divided by the funds remaining.
	$TCPI = (BAC - EV)/(BAC - AC)$ to complete on plan, or
	$TCPI = (BAC - EV)/(EAC - AC)$ to complete the current EAC.

EARNED VALUE ANALYSIS REPORT

Project Title: _____ Date Prepared: _____

Budget at Completion (BAC) _____ Overall Status _____

	Current Reporting Period	Current Period Cumulative	Past Period Cumulative
Planned value (PV)			
Earned value (EV)			
Actual cost (AC)			
Schedule variance (SV)			
Cost variance (CV)			
Schedule performance index (SPI)			
Cost performance index (CPI)			

Root Cause of Schedule Variance
Schedule Impact

EARNED VALUE ANALYSIS REPORT

Root Cause of Cost Variance

Budget Impact

	Current Reporting Period	Current Period Cumulative	Past Period Cumulative
Percent planned			
Percent earned			
Percent spent			
Estimates at Completion (EAC)			
EAC w/CPI [BAC/CPI]			
EAC w/ CPI*SPI [AC + ((BAC - EV)/ (CPI*SPI))]			
Selected EAC, Justification, and Explanation			
To complete performance index (TCPI)			

5.5 RISK REPORT

The risk report presents information on overall project risk and summarizes information on individual project risks. It provides information for identifying and analyzing risks. It also covers risk response planning, implementation, and monitoring. Typical information includes

- Executive summary
- Description of overall project risk
- Description of individual project risks
- Quantitative analysis
- Reserve status
- Risk audit results (if applicable)

The risk report can receive information from anywhere in the project environment. Some documents that should be specifically reviewed for input include

- Assumption log
- Issue log
- Lessons learned register
- Risk management plan
- Project performance reports
- Variance analysis
- Earned value status
- Risk audit
- Contractor status reports

The risk report provides information to

- Lessons learned register
- Project closeout report

The risk report is developed at the start of the project and is updated throughout the project.

Tailoring Tips

Consider the following tips to help tailor the risk report to meet your needs:

- For a small, simple, or short-term project, you can summarize this information in the regular project status report rather than create a separate risk report.
- Many projects do not include a quantitative risk analysis; if yours does not, omit this information from the report.
- For larger, longer, and more complex projects, you can tailor the quantitative risk analysis techniques used to those most appropriate to your project.
- For more robust risk reports, include appendices that may include the full risk register and quantitative risk model input (probabilistic distributions, branch correlation groups, etc.).

Alignment

The risk report should be aligned and consistent with the following documents:

- Assumption log
- Issue register
- Project performance report
- Risk management plan
- Risk register

Description

You can use the element descriptions in Table 5.5 to assist you in developing the risk report.

TABLE 5.5   Elements of a Risk Report

Document Element	Description
Executive summary	A statement describing the overall project risk exposure and major individual risks affecting the project, along with the proposed responses for trends
Overall project risk	Provide a description of the overall risk of the project, including: High-level statement of trends Significant drivers of overall risk Recommended responses to overall risk
Individual project risks	Analyze and summarize information associated with individual project risks, including: Number of risks in each box of the probability impact matrix Key metrics Active risks Newly closed risks Risks distribution by category, objective, and score Most-critical risks and changes since last report Recommended responses to top risks
Quantitative analysis	Summarize the results of quantitative risk analysis, including: Results from quantitative assessments (S-curve, tornado, etc.) Probability of meeting key project objectives Drivers of cost and schedule outcomes Proposed responses
Reserve status	Describe the reserve status, such as reserve used, reserve remaining, and an assessment of the adequacy of the reserve.
Risk audit results (if applicable)	Summarize the results of a risk audit of the risk management processes.

RISK REPORT

Project Title: _____ **Date:** _____

Executive Summary

```

```

Overall Risk Status and Trends

```

```

Significant Drivers of Overall Risk	Recommended Responses

INDIVIDUAL PROJECT RISKS

Indicate the number of individual risks in each box below.

VH					
H					
M					
L					
VL					
	VL	L	M	H	VH

RISK REPORT

Metrics

Number of scope risks	
Number of schedule risks	
Number of cost risks	
Number of quality risks	
Number of very high probability risks	
Number of high-probability risks	
Number of medium-probability risks	
Number of active risks	
Newly closed risks	

Critical Risks

Top Risks	Responses
1.	1.
2.	2.
3.	2.
4.	4.

Changes to Critical Risks

RISK REPORT

Quantitative Analysis Summary

Probability of Meeting Objectives

Scope	Schedule	Cost	Quality	Other

Range of Outcomes

Range of Schedule Outcomes	Range of Cost Outcomes

Key Drivers of Variances	Proposed Responses

Reserve Status

Total Cost Reserve	Used to Date	Used This Period	Remaining Reserve

RISK REPORT

Total Schedule Reserve	Used to Date	Used This Period	Remaining Reserve

Assessment of Reserve Adequacy

Risk Audit Summary

Summary of Risk Events

Summary of Risk Management Processes

Summary of Recommendations

5.6 CONTRACTOR STATUS REPORT

The contractor status report is filled out by the contractor and submitted on a regular basis to the project manager. It tracks status for the current reporting period and provides forecasts for future reporting periods. The report also gathers information on new risks, disputes, and issues. Information can include

- Scope performance
- Quality performance
- Schedule performance
- Cost performance
- Forecasted performance
- Claims or disputes
- Risks
- Issues
- Preventive or corrective actions

This information is generally included in the project status report compiled by the project manager. The contractor status report is submitted at predefined intervals throughout the project.

Tailoring Tips

Consider the following tips to help tailor the contractor status report to meet your needs:

- You can combine scope and quality performance into one category of technical performance.
- You can add a field for escalations to identify those areas that need to be escalated to the sponsor, program manager, contracting officer, or other appropriate individuals.
- If there were any contract change requests that were submitted during the reporting period, summary information should be described in the contractor status report.
- In addition to tailoring the content of the contractor status report, you can tailor the presentation. Many PMOs have reporting software that transforms the data into dashboards, heat reports, stop light charts, or other representations.

Alignment

The contractor status report should be aligned and consistent with the following documents:

- Procurement management plan
- Project schedule
- Cost estimates
- Project budget
- Variance analysis
- Earned value status report
- Project status report

Description

You can use the element descriptions in Table 5.6 to assist you in developing the contractor status report.

TABLE 5.6   Elements of a Contractor Status Report

Document Element	Description
Scope performance this reporting period	Describe progress on scope made during this reporting period.
Quality performance this reporting period	Identify any quality or performance variances.
Schedule performance this reporting period	Describe whether the contract is on schedule. If ahead or behind, identify the cause of the variance.
Cost performance this reporting period	Describe whether the contract is on budget. If over or under budget, identify the cause of the variance.
Forecast performance for future reporting periods	Discuss the estimated delivery date and final cost of the contract. If the contract is a fixed price, do not enter cost forecasts.
Claims or disputes	Identify any new or resolved disputes or claims that have occurred during the current reporting period.
Risks	List any risks. Risks should also be in the risk register.
Issues	Identify any new issues that have arisen. These should also be entered in the issue log.
Planned corrective or preventive action	Identify planned corrective or preventive actions necessary to recover schedule, cost, scope, or quality variances.
Comments	Add any comments that will add relevance to the report.

CONTRACTOR STATUS REPORT

Project Title: _____ Date Prepared: _____

Vendor: _____ Contract #: _____

Scope Performance This Reporting Period

Quality Performance This Reporting Period

Schedule Performance This Reporting Period

Cost Performance This Reporting Period

Forecast Performance for Future Reporting Periods

CONTRACTOR STATUS REPORT

Claims or Disputes

Risks

Planned Corrective or Preventive Action

Issues

Comments

5.7 CONTRACT CLOSEOUT REPORT

Contract closeout involves documenting the vendor performance so that the information can be used to evaluate the vendor for future work. Contract closure helps ensure contractual agreements are completed or terminated. Before a contract can be fully closed or terminated, all disputes must be resolved, the product or result must be accepted, and the final payments must be made. Information recorded as part of closing out a contract includes

- Vendor performance analysis
  - Scope
  - Quality
  - Schedule
  - Cost
  - Other information, such as how easy the vendor was to work with
- Record of contract changes
  - Change ID
  - Description of change
  - Date approved
- Record of contract disputes
  - Description of dispute
  - Resolution
  - Date resolved

The date of contract completion, who signed off on it, and the date of the final payment are other elements that should be recorded.

Tailoring Tips

Consider the following tips to help tailor the contract closeout report to meet your needs:

- For a small contract, you can combine all the information in a vendor performance analysis into a summary paragraph.
- For small contracts, you may not need information on contract changes or contract disputes.
- If the project was based around one large contract, you can combine the information in the project closeout report with this form.

Alignment

The contract closeout report should be aligned and consistent with the following documents:

- Procurement management plan
- Procurement audit
- Change log
- Project closeout report

Description

You can use the element descriptions in Table 5.7 to assist you in developing the contract closeout report.

TABLE 5.7   Elements of a Contract Closeout

Document Element	Description	
What worked well	Scope	Describe aspects of contract scope that were handled well.
	Quality	Describe aspects of product quality that were handled well.
	Schedule	Describe aspects of the contract schedule that were handled well.
	Cost	Describe aspects of the contract budget that were handled well.
	Other	Describe any other aspects of the contract or procurement that were handled well.
What can be improved	Scope	Describe aspects of contract scope that could be improved.
	Quality	Describe aspects of product quality that could be improved.
	Schedule	Describe aspects of the contract schedule that could be improved.
	Cost	Describe aspects of the contract budget that could be improved.
	Other	Describe any other aspects of the contract or procurement that could be improved.
Record of contract changes	Change ID	Enter the change identifier from the change log.
	Change description	Enter the description from the change log.
	Date approved	Enter the date approved from the change log.
Record of contract disputes	Description	Describe the dispute or claim.
	Resolution	Describe the resolution.
	Date resolved	Enter the date the dispute or claim was resolved.

CONTRACT CLOSEOUT

Project Title: _____ **Date Prepared:** _____

Project Manager: _____ **Contract Representative:** _____

Vendor Performance Analysis

What Worked Well	
Scope	
Quality	
Schedule	
Cost	
Other	
What Can Be Improved	
Scope	
Quality	
Schedule	
Cost	
Other	

Record of Contract Changes

Change ID	Change Description	Date Approved

CONTRACT CLOSEOUT

Record of Contract Disputes

Description	Resolution	Date Resolved

Date of Contract Completion _____

Signed Off by _____

Date of Final Payment _____

5.8 LESSONS LEARNED REPORT

Lessons learned are compiled throughout the project or at specific intervals, such as at the end of a life cycle phase. These are recorded in the lessons learned register. The lessons learned summary compiles and organizes those things that the project team did that worked very well and should be passed along to other project teams and identifies those things that should be improved for future project work. The summary should include information on risks, issues, procurements, quality defects, and any areas of poor or outstanding performance. Information that can be documented includes

- Project performance analysis
  - Requirements
  - Scope
  - Schedule
  - Cost
  - Quality
  - Physical resources
  - Team development and management
  - Communication
  - Risk management
  - Procurement management
  - Stakeholder management
  - Process improvement
  - Product-specific information
- Information on specific risks
- Quality defects
- Vendor management
- Areas of exceptional performance
- Areas for improvement

This information is saved, along with the lessons learned register, in a lessons learned repository. Repositories can be as simple as a lessons learned binder, they can be a searchable database, or anything in between. The purpose is to improve performance on the current project (if done during the project) and future projects. Use the information from your project to tailor the form to best meet your needs. The lessons learned summary is developed at the close of a phase for long projects and at the close of the project for shorter projects.

Tailoring Tips

Consider the following tips to help tailor the lessons learned summary to meet your needs:

- Add, combine, or eliminate rows as needed to capture the important aspects of your project.
- Consider including a section on change management, as that can sometimes be a challenging aspect of projects to manage.
- You may want to include a section on phase management if you are doing the summary at the end of the project.
- If your project used a new development approach, such as a blended predictive and adaptive (agile) approach, consider adding some relevant lessons learned.

Alignment

The lessons learned summary should be aligned and consistent with the following documents:

- Issue register
- Risk register
- Decision log
- Lessons learned register
- Retrospectives

Description

You can use the element descriptions in Table 5.8 to assist you in developing the lessons learned summary.

TABLE 5.8 Elements of a Lessons Learned Summary

Document Element	Description	
Project Performance	**What Worked Well**	**What Can Be Improved**
Requirements definition and management	List any practices or incidents that were effective in defining and managing requirements.	List any practices or incidents that can be improved in defining and managing requirements.
Scope definition and management	List any practices or incidents that were effective in defining and managing scope.	List any practices or incidents that can be improved in defining and managing scope.
Schedule development and control	List any practices or incidents that were effective in developing and controlling the schedule.	List any practices or incidents that can be improved in developing and controlling the schedule.
Cost estimating and control	List any practices or incidents that were effective in developing estimates and controlling costs.	List any practices or incidents that can be improved in developing estimates and controlling costs.
Quality planning and control	List any practices or incidents that were effective in planning, managing, and controlling quality.	List any practices or incidents that can be improved in planning, managing, and controlling quality. Specific defects are addressed elsewhere.
Physical resource planning and control	List any practices or incidents that were effective in planning, acquiring, and managing physical resources.	List any practices or incidents that can be improved in planning, acquiring, and managing physical resources.
Team, development, and performance	List any practices or incidents that were effective in working with team members and developing and managing the team.	List any practices or incidents that can be improved in working with team members and developing and managing the team.
Communications management	List any practices or incidents that were effective in planning and distributing information.	List any practices or incidents that can be improved in planning and distributing information.
Risk management	List any practices or incidents that were effective in the risk management process. Specific risks are addressed elsewhere.	List any practices or incidents that can be improved in the risk management process. Specific risks are addressed elsewhere.

TABLE 5.8   Elements of a Lessons Learned Summary (*continued*)

Document Element	Description	
Procurement planning and management	List any practices or incidents that were effective in planning, conducting, and administering contracts.	List any practices or incidents that can be improved in planning, conducting, and administering contracts.
Stakeholder engagement	List any practices or incidents that were effective in engaging stakeholders.	List any practices or incidents that can be improved in engaging stakeholders.
Process improvement information	List any processes that were developed that should be continued.	List any processes that should be changed or discontinued.
Product-specific information	List any practices or incidents that were effective in delivering the specific product, service, or result.	List any practices or incidents that can be improved in delivering the specific product, service, or result.
Other	List any other practices or incidents that were effective, such as change control, configuration management, etc.	List any other practices or incidents that can be improved, such as change control, configuration management, etc.
Risks and issues	Risk or issue description	Identify risks or issues that occurred that should be considered to improve organizational learning.
	Response	Describe the response and its effectiveness.
	Comments	Provide any additional information needed to improve future project performance.
Quality defects	Defect description	Describe quality defects that should be considered to improve organizational effectiveness.
	Resolution	Describe how the defects were resolved.
	Comments	Indicate what should be done to improve future project performance.
Vendor management	Vendor	List the vendor(s).
	Issue	Describe any issues, claims, or disputes that occurred.
	Resolution	Describe the outcome or resolution.
	Comments	Indicate what should be done to improve future vendor management performance.
Other	Areas of exceptional performance	Identify areas of exceptional performance that can be passed on to other teams.
	Areas for improvement	Identify areas that can be improved on for future performance.

LESSONS LEARNED REPORT

Project Title: _____ **Date Prepared:** _____

Project Performance Analysis

	What Worked Well	What Can Be Improved
Requirements definition and management		
Scope definition and management		
Schedule development and control		
Cost estimating and control		
Quality planning and control		
Physical resource planning and control		
Team development and performance		
Communications management		
Risk management		
Procurement planning and management		
Stakeholder engagement		
Process improvement information		
Product-specific information		
Other		

Risks and Issues

Risk or Issue Description	Response	Comments

LESSONS LEARNED REPORT

Quality Defects

Defect Description	Resolution	Comments

Vendor Management

Vendor	Issue	Resolution	Comments

Other

Areas of Exceptional Performance	Areas for Improvement

5.9 PROJECT CLOSEOUT REPORT

Project closeout involves documenting the final project performance as compared to the project objectives. The objectives from the project charter are reviewed and evidence of meeting them is documented. If an objective was not met, or if there is a variance, that is documented as well. In addition, information from the procurement closeout is documented. Information documented includes

- Project description
- Project objectives
- Completion criteria
- How met
- Cost and schedule variances
- Benefits management
- Business needs
- Summary of risks and issues

Use the information from your project to determine the best approach.

Tailoring Tips

Consider the following tips to help tailor the project or phase closeout to meet your needs:

- For longer projects, consider a phase closeout rather than waiting until the end of the project. At the end of the project, you can compile all the phase closeout information.
- When working with an incremental life cycle or Agile development method, the delivery of a major end item, service, or capability may benefit from a formal phase closeout report.
- For projects that are part of a program, you should tailor the content to meet the needs of the program.

Alignment

The project closeout should be aligned and consistent with the following documents:

- Project management plan (all components)
- Lessons learned summary

The project closeout form is developed at the end of a project or phase.

Description

You can use the element descriptions in Table 5.9 to assist you in developing the project or phase closeout.

TABLE 5.9 Elements of a Project or Phase Closeout

Document Element	Description	
Project description	Provide a summary level description of the project.	
Performance summary	Scope	Describe the scope objectives needed to achieve the planned benefits of the project.
		Document the specific and measurable criteria needed to complete the scope objectives.
		Provide evidence that the completion criteria were met.
	Quality	Describe the quality objectives and criteria needed to achieve the planned benefits of the project.
		Document the specific and measurable criteria needed to meet the product and project quality objectives.
		Enter the verification and validation information from the product acceptance form.
Variances	Document the time and cost objectives and the final completion date and final expenditures. Explain any variances.	
Business needs	Describe how the final product, service, or result achieved the business needs identified in the business plan.	
Risks and issues	Summarize any significant risks or issues, or the overall risk exposure, and describe the response and resolution strategies.	

PROJECT CLOSEOUT

Project Title: **Date:**   **Project Manager:**

Project Description

Performance Summary

	Objectives	Completion Criteria	How Met
Scope			
Quality			

Variances

	Objectives/Final Outcome	Variances	Comments
Time			
Cost			

Business Needs

PROJECT CLOSEOUT

Risks and Issues

Risk or Issue	Response or Resolution	Comments

5.10 QUALITY AUDIT

A quality audit is a technique that employs a structured, independent review to project and/or product elements. Any aspect of the project or product can be audited. Common areas for audit include

- Project processes
- Project documents
- Product requirements
- Product documentation
- Defect or deficiency repair
- Compliance with organizational policies and procedures
- Compliance with the quality management plan
- Good practices from similar projects
- Areas for improvement
- Description of deficiencies or defects

Defects or deficiencies should include action items, a responsible party, and be assigned a due date for compliance.

A quality audit is conducted at predetermined intervals, or as needed.

Tailoring Tips

Consider the following tips to help tailor the quality audit to meet your needs:

- Quality audits can also include information that will be shared with other projects.
- Some projects use audits to track the implementation of approved changes and corrective or preventive actions.

Alignment

The quality audit should be aligned and consistent with the following document:

- Quality management plan

Description

You can use the element descriptions in Table 5.10 to assist you in developing the quality audit.

TABLE 5.10   Elements of a Quality Audit

Document Element	Description
Area audited	Check the box for the area or areas audited.
Good practices from similar projects	Describe any good or best practices that can be shared from similar projects.
Areas for improvement	Describe any areas that need improvement and the specific improvements or measurements that need to be achieved.

TABLE 5.10   Elements of a Quality Audit (*continued*)

Document Element	Description	
Deficiencies or defects	ID	Enter a unique defect identifier.
	Defect	Describe the deficiency or defect.
	Action	Describe the corrective actions needed to fix the defect.
	Responsible party	Identify the person assigned to correct the deficiency or defect.
	Due date	Document the due date.
Comments	Provide any additional useful comments about the audit.	

QUALITY AUDIT

Project Title: _____ **Date Prepared:** _____

Project Auditor: _____ **Audit Date:** _____

Area Audited:	
☐ Project processes	☐ Project documents
☐ Product documents	☐ Product documentation
☐ Quality management plan	☐ Defect/deficiency repair
☐ Organizational policies and procedures	

Good Practices from Similar Projects

Areas for Improvement

Deficiencies or Defects				
ID	Defect	Action	Responsible Party	Due Date

Comments

5.11 RISK AUDIT

Risk audits are used to evaluate the effectiveness of the risk identification, risk responses, and risk management process as a whole. Information reviewed in a risk audit can include

- Risk event audits
  - Risk events
  - Causes
  - Responses
- Risk response audits
  - Risk event
  - Responses
  - Success
  - Actions for improvement
- Risk management processes
  - Process
  - Compliance
  - Tools and techniques used
- Good practices
- Areas for improvement

The risk audit is conducted periodically as needed.

Tailoring Tips

Consider the following tips to help tailor the risk audit to meet your needs:

- To make the audit more robust, you can include an assessment of the effectiveness of the risk management approach.
- Large organizations often have policies and procedures associated with project risk management. If this is the case in your organization, include an assessment of compliance with the policies and procedures.
- Many organizations don't track opportunity management. You can expand the scope of the audit to include opportunity management if appropriate.
- For larger projects, you may want to include information on overall risk in addition to risk events.

Alignment

The risk audit should be aligned and consistent with the following documents:

- Risk management plan
- Risk register
- Risk report

Description

You can use the element descriptions in Table 5.11 to assist you in developing the risk audit.

TABLE 5.11   Elements of Risk Audit

Document Element	Description	
Risk event audit	Event	List the event from the risk register.
	Cause	Identify the root cause of the event from the risk register.
	Response	Describe the response implemented.
	Comment	Discuss if there was any way to have foreseen the event and respond to it more effectively.
Risk response audit	Event	List the event from the risk register.
	Response	List the risk response from the risk register.
	Successful	Indicate if the response was successful.
	Actions to improve	Identify any opportunities for improvement in risk response.
Risk management process audit	Risk management planning	Followed: Indicate if the various processes were followed as indicated in the risk management plan.
	Risk identification	Tools and techniques used: Identify tools and techniques used in the various risk management processes and whether they were successful.
	Risk analysis	
	Risk response Planning	
	Risk monitoring	
Description of good practices to share	Describe any practices that should be shared for use on other projects. Include any recommendations to update and improve risk forms, templates, policies, procedures, or processes to ensure these practices are repeatable.	
Description of areas for improvement	Describe any practices that need improvement, the improvement plan, and any follow-up dates or information for corrective action.	

RISK AUDIT

Project Title: _____ Date Prepared: _____

Project Auditor _____ Audit Date _____

Risk Event Audit

Event	Cause	Response	Comment

Risk Response Audit

Event	Response	Successful	Actions to Improve

Risk Management Process Audit

Process	Followed	Tools and Techniques Used
Risk Management Planning		
Risk Identification		
Risk Analysis		
Risk Response Planning		
Risk Monitoring		

RISK AUDIT

Description of Good Practices to Share

Description of Areas for Improvement

5.12 PROCUREMENT AUDIT

A procurement audit reviews contracts and contracting processes for completeness, accuracy, and effectiveness. Information in the audit can be used to improve the process and results on the current procurement and on other contracts. Information recorded in the audit includes

- Vendor performance audit
  - ○ Scope
  - ○ Quality
  - ○ Schedule
  - ○ Cost
  - ○ Other information
- Procurement management process audit
  - ○ Process
  - ○ Tools and techniques used
- Description of good practices
- Description of areas for improvement

The procurement audit is conducted periodically throughout the project, or as needed.

Tailoring Tips

Consider the following tips to help tailor the procurement audit to meet your needs:

- Add qualitative information, such as how easy the vendor was to work with, timeliness of returning calls, and collaborative attitude. This can provide useful information for future procurement opportunities.

Alignment

The procurement audit should be aligned and consistent with the following documents:

- Procurement management plan
- Contractor status report
- Contract closeout report

Description

You can use the element descriptions in Table 5.12 to assist you in developing the procurement audit.

TABLE 5.12   Elements of a Procurement Audit

Document Element	Description	
What worked well	Scope	Describe aspects of contract scope that were handled well.
	Quality	Describe aspects of product quality that were handled well.
	Schedule	Describe aspects of the contract schedule that were handled well.
	Cost	Describe aspects of the contract budget that were handled well.

(continued)

TABLE 5.12 Elements of a Procurement Audit (*continued*)

Document Element	Description		
	Other	Describe any other aspects of the contract or procurement that were handled well.	
What can be improved	Scope	Describe aspects of contract scope that could be improved.	
	Quality	Describe aspects of product quality that could be improved.	
	Schedule	Describe aspects of the contract schedule that could be improved.	
	Cost	Describe aspects of the contract budget that could be improved.	
	Other	Describe any other aspects of the contract or procurement that could be improved.	
Procurement management process audit	Procurement Planning	Indicate if each procurement was followed or not.	Describe any tools or techniques that were effective for each procurement.
	Conducting procurements		
	Procurement Management		
Good practices to share	Describe any good practices that can be shared with other projects or that should be incorporated into organization policies, procedures, or processes. Include information on lessons learned.		
Areas for improvement	Describe any areas that should be improved with the procurement process. Include information that should be incorporated into policies, procedures, or processes. Include information on lessons learned.		

PROCUREMENT AUDIT

Project Title: _____ Date Prepared: _____

Project Auditor: _____ Audit Date: _____

Vendor Performance Audit

What Worked Well
Scope
Quality
Schedule
Cost
Other
What Can Be Improved
Scope
Quality
Schedule
Cost
Other

Procurement Management Process Audit

Process	Followed	Tools and Techniques Used
Procurement Planning		
Conducting Procurements		
Managing Procurements		

PROCUREMENT AUDIT

Description of Good Practices to Share

Description of Areas for Improvement

Appendix: Combination Templates

Throughout the book, I have mentioned that you can combine certain templates if your project warrants it. This appendix provides an example of five combined templates:

- Vision statement + project proposal
- Charter + scope statement
- Scope management plan + quality management plan
- Communications management plan + stakeholder engagement plan
- Procurement strategy + source selection

Because all the content in these templates is discussed in the book, I will not repeat it here. However, the table below gives a brief summary of how I tailored the templates when I combined them.

Documents	Tailoring Description
Vision statement Project proposal	This is simply a combination of the two templates. All content from each template is included.
Charter Scope statement	This is an abbreviated charter with scope statement information integrated into it. Information that is eliminated from the charter includes high-level requirements, overall project risk, stakeholders, project exit criteria, and project manager authority levels. Information from the scope statement that is part of the abbreviated charter includes exclusions and deliverable acceptance criteria.
Scope management plan Quality management plan	This template takes an abbreviated version of each template and combines them. The template contains information about decomposing and organizing scope, which can refer to waterfall or Agile methods. The quality objectives are identified by listing a deliverable, metric, measure, and acceptance criteria. There is a place to describe which deliverables and processes are subject to quality review and a place to describe how scope change management should take place.

Documents	Tailoring Description
communications management plan Stakeholder engagement plan	This template takes an abbreviated version of each template and combines them. Information on the sender, assumptions, constraints, and glossary of terms is eliminated from the communications management plan. Information on the current and desired level of engagement, pending stakeholder changes, and stakeholder relationships is eliminated from the stakeholder engagement plan. There are two new fields that were added, one for influence and one for attitude. Influence refers to the degree to which a stakeholder can influence the project direction or decisions. Attitude refers to whether a stakeholder feels favorably or unfavorably about the project.
Procurement strategy Source selection	This is simply a combination of the two templates. All content from each template is included.

These template examples are merely a few options for combining and tailoring the templates in this book. Perhaps looking at how these templates are modified and combined will give you some ideas on how to tailor templates for your use

PROJECT PROPOSAL

Proposed Project Title: _____ **Date:** _____

Executive Summary:

```
                                                                      
                                                                      
                                                                      
                                                                      
                                                                      
                                                                      
```

Project Vision

We are developing _____ for _____.

To respond to the following need(s):

-
-
-

This product responds to those needs by providing the following key attributes:

-
-
-
-

Customers will buy this product because of these benefits:

-
-
-
-

Project Background:

```
                                                                      
                                                                      
                                                                      
                                                                      
```

PROJECT PROPOSAL

Solution and Approach:

Goals	Scope

Financial Information

Resource Requirements

Physical Resources	Team Resources

Conclusion

ABBREVIATED PROJECT CHARTER

Project Title: _____ Date Prepared: _____

Project Sponsor: _____ Project Manager: _____

Project Purpose

High-Level Project Description

Project Boundaries and Exclusions

Project Deliverables	Acceptance Criteria

ABBREVIATED PROJECT CHARTER

	Project Objectives	Success Criteria
Scope		
Time		
Cost		
Other		

Summary Milestones	Due Date

Budget

Approvals

Project Manager Signature	Sponsor or Originator Signature
Project Manager Name	Sponsor or Originator Name
Date	Date

SCOPE MANAGEMENT PLAN

Project Title: _____ Date: _____

Decomposing and Organizing Scope

Quality Objectives

Deliverable	Metric or Specification	Measure	Acceptance Criteria

Deliverables and Processes Subject to Quality Review

Deliverables	Processes

Scope Change Management

COMMUNICATIONS MANAGEMENT PLAN

Project Title: _____   **Date Prepared:** _____

Stakeholder	Influence	Attitude	Information	Method	Timing or Frequency

Stakeholder Engagement Approach

Stakeholder	Approach

PROCUREMENT STRATEGY

Project Title: _____   **Date:** _____

Delivery Method

Contract Type

☐ FFP	☐ FPIF	☐ FP-EPA	☐ CPFF	☐ CPIF	☐ CPAF	☐ T&M	☐ Other

Incentive or Award Fee	Criteria

Procurement Life Cycle

Phase	Entry Criteria	Key Deliverables or Milestones	Exit Criteria	Knowledge Transfer

PROCUREMENT STRATEGY

Source Selection

Score	1	2	3	4	5
Criterion 1					
Criterion 2					
Criterion 3					

	Weight	Candidate 1 Rating	Candidate 1 Score	Candidate 2 Rating	Candidate 2 Score	Candidate 3 Rating	Candidate 3 Score
Criterion 1							
Criterion 2							
Criterion 3							
Totals							

Index